Risk Controlling In Stochastic Optimization Of EVA Architecture

Neil Boron

Chapter 1
INTRODUCTION

1.1 Background

Significant changes have occurred all over the world even over the past decade in the energy landscape. Globally, there's a big push towards a 100% incorporation of wind and solar power for electricity production, with synergistic support from Vehicle-to-Grid (V2G) technology. Large-scale electric vehicles (EVs) can perform as new channels of source and energy storage mechanism in the smart grid via implementing V2G intended for power injection from EVs to the grid. V2G awards EV owners with numerous innovative roles in the smart grid such as mobile energy storage, renewable energy storage, backup supply. V2G has been intended as a means of lessening the grid potential operational delinquents and enhancing the opportunity for the proliferation of EVs. There are however a host of challenges, most of which are due to the intermittency and unpredictability of the mobility behaviour and renewable energy resources (RESs), along with market price volatility. Therefore, V2G scheduling becomes emerging research focus with a large-scale application of EVs. Present market regulations restrict fair participation in power markets for individual EVs' G2V scheduling. The complete financial value of EVs would be realized provided policy outlines, market rules and regulatory reforms allow EV Aggregator (EVA) to provide multiple services (energy and ancillary services) simultaneously, addressing congestion at the system level, paying attention to satisfying reserve requirements, and serving to accomplish energy management for single EV owner. It enables EVA to ensure that EVs can receive fair remuneration for unlocking their multi-valued streams and providing flexibility services in a cost-competitive manner. If aggregated and controlled, EVs can provide flexibility to the system operator (SO). This thesis will focus on some of the possible solutions for the V2G integration for providing ancillary services that are control-centric. This thesis will cover some of these challenges to V2G scheduling, highlights the current research in V2G optimization, and use-case studies that illustrate the role of EVA in power grids and electricity markets. Smart V2G scheduling can reshape the load profile and further decrease substantial investment costs and minimalize operative costs, by adopting G2V mode in off-peak duration and V2G mode when demand is high. EVA can maximize their flexibility provision from both the demand side (G2V) and supply-side (V2G).

1) G2V mode: EVs are functioned as a demand-side managing alternative, forming a flexible electricity consumption supporting renewable power integration.

2) V2G mode: EVs are operated as dispersed generation resources. The batteries of EVs could inject power to the grid by discharging and forming large-scale distributed electricity storage regulated/controlled by the needs of an electric system. V2G control might offer an important revenue stream that would enhance the economics of grid-integrated EVs and additionally motivate their adoption. It could also recuperate grid stability. However, the downside of V2G operation is namely, reduced battery life, additional infrastructure cost, safety issues, etc.

Utilizing a mobile energy storage system, EV batteries may store electric power which may assist as a grid source during peak periods. Employing distributed energy storage of EVs, EVAs could offer ancillary services *viz.* spinning reserves and frequency regulation. EV owners can be incentivized for motivation towards offering V2G services. In other words, their market participation can be facilitated to perform charging operation whenever the network experiences less demand and execute discharging at times when there is most demand on the network. Nevertheless, the V2G market entails strict regulations and integration of demand response prioritize are desired to enhance the market efficiency of V2G scheduling. The dynamics of mobility behaviour of EV owners elevates challenges to V2G scheduling of EVA.

The EV batteries usage standards are very strict from their state of health aspect. Flexibility services use EV's manageable charging/discharging capability to adapt power rates at the network contact notch. Conventionally this servicing is all activities near elementary charging (whose intention is to upsurge the SOC to and to guarantee energy for driving) and they can be: Regulation-Up attained by decrease/increase of charging/ discharging power in charging G2V/ discharging V2G mode of operation, and start discharging if in idle mode. Regulation-Down realized by increase/decrease of charging/ discharging power, in charging G2V/ discharging V2G mode of operation, and start charging if in idle mode.

1.2 The Motivation for Present Work

In the gleam of the low-carbon economy, EVs are an attractive and vital instrument to bond the gap bounded by conventional and low-carbon power networks. The EV is premeditated globally as a propitious option to traditional internal combustion engine vehicles since it can lessen the GHG emissions from the transportation sector and employ surplus electric power from RES as well. It has been assessed that the entire EVs' uncontrolled charging demand in the United States (US) can spread 18% of the US summer peak at 30% EV integration. EVs' uncoordinated charging might endanger the reliability and security of the distribution network, especially in terms of peak load.

The increasing penetrations of renewable energy in the electricity sector and EVs in the transportation sector have increased the interests in introducing new methods to deal with uncertainties in power system studies. Ancillary services are provided to help with grid stability and reliability. Ancillary services are grid support services essential for the SOs to sustain the integrity, stability and power quality of the distribution system. Ancillary services are customized as per requirements in Grid Codes and few services are procured as desired by SOs to retain the frequency and voltage of the distribution system within operative parameters or to recuperate the system under disturbance or failure. Present market regulations restrict participation in power markets for individual EVs requiring EVA for facilitating their market participation.

Unregulated charging is the baseline scenario of charging whenever and wherever parked, i.e. an EV is charged after it is plugged in on arrival at home/a charge station. The EV keeps charging up until the battery box is completely charged or unplugged for driving purpose. Concurrent charging case, live typically at peak duration, surges the charging cost burden on EV owners. The increased variation of charging rate could speed up battery degradation. Thus, unregulated EVs' charging (concentrated charging, along with an increased variety of charging rate) would posture major threats in electricity markets and system/grid operations. It may increase peak loading and necessitates network reinforcement which is not cost-effective. Coordinated G2V charge scheduling is required to mitigate the negative impacts of increased network loading. A stochastic decision-making framework is desired to deal with uncertainties faced by EVA. To incorporate the risk associated with the market price volatility and to assess the effect of scheduling energy and regulation on the profit variability, a suitable risk measure is required for V2G operation optimization through orchestration of risk appetites and multiple perspectives.

Demand Response Programs (DRP) have garnered attention due to their substantial capabilities and potentials. The specially designed effective pricing structure for G2V charge scheduling by EVA establishes a critical problem on the way to the cognizance of the substantial EV flexibility potentiality in deregulated power systems. The coordination between the involved entities can bring more flexibility to enhance the total profitability by distributing charging over the plug-in duration. It necessitates a scalable G2V scheduling method capturing the uncertain mobility behaviour of EV owners, outperforming about solution improvement and information handing out necessities, and extensively evaluate and explore the monetary influences of synergized flexibility and smart- G2V scheduling performance on SO, EVA and EV owners. The goal is a new operational model using Price-based Demand Response (PBDR) that will assist economic development in V2G scheduling. Particular focus will be put on how PBDR could be intelligently orchestrated to provide grid flexibility services and participate in different energy markets, as well as autonomously controlled to provide regulation services for supporting grid stability. The mission of the present study is to assess the economics of risk-averse G2V scheduling of EVA for multiple markets participation and enhance its market participation efficiency through the integration of PBDR. The final aim is to illustrate how the integration of PBDR for EVA's V2G scheduling can facilitate an affordable, secure and resilient transition towards a low-carbon energy future.

1.3 Contributions of the Present Work

The purpose of the present research work includes the study of fundamentals of integrated Time-of-Use (TOU) -PBDR and risk-averse V2G scheduling of EVA for supporting grid stability and improved market operations through a worldwide literature survey. The main contributions of the present research are outlined as:

1. A detailed state-of-art survey related to smart V2G scheduling strategies for EVA and methodologies in practice has been made. This further includes the latest charging standards, best markets for their participation, balancing of techno-economic concerns, challenges for designing cost-effective V2G scheduling from different perspectives of entities involved and dynamics of methodology. Based on the literature survey on V2G scheduling, an overview of the research problem is presented. The in-depth study on various aspects helps to understand the requirement of deferring network reinforcement through enhancements in the existing V2G scheduling framework with due consideration of evolving low carbon scenario, market price volatility and mobility behaviour dynamics.

2. Individual EV owners can't regulate their charging most efficiently. Therefore, EVA performs as an interface between SO and EV owners to facilitate their market participation for providing regulation services to SO. Considering EV owner perspective to reduce charging cost, a part of this thesis proposes static G2V scheduling of EVA with complete charging profile information to support grid stability through smart coordination among them and outperforms as compared to uncoordinated scheduling. The sensitivity analysis of EVA's performance parameters: revenue, cost and profit are accomplished concerning the number of EVs and upper limit of charging rate to validate the effectiveness of the proposed approach. Optimal solutions specified by static G2V charge scheduling approaches could be employed to illustrate the possible advantages that could be delivered by regulated charge scheduling and provide a benchmark for performance assessment.
3. EVA adopts harmonized and dispersed charging rates for EVs by the plugged-in duration offering regulation and stability to SO. Nevertheless, static G2V scheduling is not a realistic case. Therefore, the contribution of this thesis is to propose a more pragmatic dynamic G2V scheduling of EVA. Scenarios of times of arrival & departure and initial SOC are generated availing the Monte Carlo Simulation (MCS) method. The probabilistic Kantorovich Distance (KD) is adopted in the scenario reduction backward algorithm to arbitrate a trade-off between computation speed and accuracy. Furthermore, RTPDR, frequently endorsed in literature, has a low acceptance rate by EV owners for being too unstable to respond and being not viable for a few charge cycles in a day. TOU, being less dynamic, is a natural price signal under PBDR to EV owners for motivating them to operate in G2V mode when the network experiences low demand and employ their flexibility for load flattening and peak alleviation. TOU-PBDR decreases EV charging price as well. A contribution of this thesis is to propose a TOU-PBDR incorporated framework for EVA's G2V charge scheduling for the energy market and regulation services. EV owner viewpoint for charging cost reduction is deliberated accompanied by EVA profit maximization. TOU is drafted from RTP employing the agglomerative hierarchical clustering (AHC) method. EVs mobility ambiguity characterization causes scheduling dynamic. Sensitivity analysis concerning higher bound of charging rate and numeral of EVs validates the method. Results demonstrate

fascinating directions respecting EV charging cost, EVA profit, and regulation provision to SO.

4. In a smart grid system, the price volatility would necessarily impact market transactions and the behaviour of an EVA intending profit escalation through G2V scheduling. EVA could motivate EV owners for G2V mode during off-peak times by incorporating PBDR in dynamic G2V operational scheduling and employ their flexibility for load flattening, peak alleviation, and reducing the charging payment as well. For enhanced multi-market partaking, risk-averse G2Vscheduling of EVA is modelled in the present thesis to manage its revenue risk and decrease energy procurement cost integrating TOU-PBDR. Similar to the previous chapter stochasticity of prices is handled through the Monte Carlo Simulation (MCS) method and the probabilistic Kantorovich Distance (KD) approach. Risk measure unified is the Conditional Value at Risk ($CVaR$). Simulation results validate the effectiveness of the proposed scheme and the impact of the stochastic G2V scheduling plan of EVA on the energy market and regulation operations.
5. EVs dynamic mobility behaviour and market price volatility as vital sources of multiple uncertainties, introduces the risk to EVA, affecting severely its G2V operational behaviour. As mentioned earlier, TOU-PBDR being less fluctuating is an accepted price indicator with a high compliance rate by EV owners but might lower EVA's profit owing to market price volatility. Considering this, the final contribution of this thesis is to propose a risk-averse integrated TOU-PBDR framework of EVA to control its revenue risk and lessen energy procurement cost, meeting schedule plans and managing the adverse effect of multiple uncertainties: market price and times of arrival and departure. The effect of these uncertainties on the G2V scheduling is characterized through a stochastic programming architecture. Risk measure incorporated is $CVaR$. The outcomes from prudent case studies illustrate that perspicacity depends upon the anticipated method offering improved trade-off concerning profit and risk.
6. Unidirectional G2V has some restrictions such as the regulation capacities offer with G2V are substantially lower than those that can be offered in bidirectional V2G. G2V cannot deliver the SO with EVs' stored energy as well. V2G, the provision of energy and ancillary services from an EV to the grid, can provide techno-economic benefits to SO and EV owners. In this work, risk-aversive

integrated DR and V2G scheduling algorithm are developed to optimize the decision-making concerning multi-market (viz. energy and regulation services) participation of EVA. EVA utilizes this developed algorithm to maximize its expected profits at trade-off risk from uncertain mobility behaviour and market prices while reducing the charging costs of EV owners and offer further flexibility and peak shaving to the SO, which can aid the existing distribution system in integrating large-scale intermittent renewable energy sources. The presented approach sets charging/discharging rates based on offered TOU price for each EV owner emphasizing its battery health perspective. Sensitivity analysis of performance parameters concerning battery replacement cost and round-trip efficiency is analysed. The obtained results show that consideration of the degradation model decreases the obtained profit in the short-term run. Furthermore, a risk management technique, using coherent risk measure CVaR is executed and compared with the risk-neutral problem. The expected profit for risk-neutral, RTP-DR, utility designed TOU-DR and designed TOU-DR are compared. Considering TOU-PBDR from a customer perspective, the expected profit in designed TOU-DR is obtained closer to RTP-DR whereas utility TOU-DR reduced the expected profit of EVA meaningfully. The risk-averse leads to a more conservatism condition compared to risk-neutral case, and the traded energy is reduced significantly. Simulation results demonstrate the rationality of this method to maintain an equilibrium among different aspects of involved entities. As a part of future work, integrated DR and risk-aversive V2G formulation can be executed for better decision-making by EVA in multiple markets considering balance among the preferences of involved entities and large-scale renewable integration.

1.4 Organization of the Thesis

This doctoral thesis started with an overview of the G2V/V2G scheduling framework of EVA and the benefits of integrating DR for its market participation efficiency enhancement. The research gap outlined in this thesis helps to realize the encouragement for research in the present field of green transportation, and analyses the current issues in the area of G2V/V2G scheduling of EVA. It further summarizes the contribution of the present thesis. This thesis has six chapters, including chapter one entitled "Introduction". The rest of the chapters of this thesis are organized as follows:

Chapter 2 offers an overview of various country practices adopted for EVA's G2V/V2G scheduling, along with underlining the necessary framework adopted. Further different roles of EVAs and their current status and participation in the energy and ancillary services markets for supporting grid stability are illustrated. Also, the main methodologies and challenges appearing in G2V/V2G scheduling of EVA are presented and the necessity of enhancement in the existing model from EV owners' perspective is also discussed.

Chapter 3 proposes different aspects related to grid integrated EVs and grid support services. These aspects include static charging method, charging strategy, sensitivity analysis of the varying number of EVs and upper limit of charging rate on performance parameters of EVA.

Chapter 4 proposes an integrated DR for improved market operations and a more realistic dynamic smart-charging algorithm. Besides, a stochastic programming approach is used to characterize uncertain mobility parameters (times of arrival and departure; and initial State-of-Charge SOC) and solve the optimization problem.

Chapter 5 elaborates the integrated DR and static G2V model presenting the risk-neutral and risk-averse behaviours of EVA considering market price uncertainties. Real consumption data of households have been used. Then, a new driving and charging behaviour model has been developed based on the 2009 National Household Travel Survey (NHTS), in which the database is freely available. Efficient frontiers are obtained as a variation of *CVaR* and expected profit for varying risk weighting parameter β.

Chapter 6 elaborates integrated DR and dynamic G2V scheduling for risk-neutral and risk-averse EVA considering multiple uncertainties: market price volatility and mobility behaviour uncertainty.

Chapter 7 develops risk-averse integrated DR and V2G scheduling algorithm of EVA in energy and ancillary service markets to maintain an equilibrium among different preferences of involved entities SO, EV owners, and EVA.

Chapter 8 concludes some major findings of the thesis. A summary of the results shown in Chapters 3-7 of the thesis is also presented. Thus, it gathers the conclusions drawn from the research work developed in this thesis and proposes future areas that could be explored to continue the research initiated in this thesis.

Chapter 2

LITERATURE REVIEW

2.1 Introduction

International Energy Agency (IEA) reports [1-5] perceives EVs as a propitious pathway in the transportation sector contributing to alleviate greenhouse gas emissions and aid reduction of mobility carbon footprint around the globe. However, large-scale integration of EVs can adversely impact grid stability and increased reserve requirements [6]. On way towards a transition to sustainable development desires cheap, reliable and clean energy through Renewable Energy Resources (RES): wind and Photovoltaic (PV) generation. RES has inherent intermittency issues requiring mobile energy storage of EVs as a cost-effective solution. Large-scale RE integration requires balancing services not only at the transmission level but also on the local level whereas massive EVs unregulated charging creates new peaks over existing peaks in evening hours causing network congestion and increased losses [7-10]. It may lead to grid instability and peak demand rise.

As an ancillary services product, the regulation provides market-based compensation to resources that can adjust output or consumption in response to an automated signal [11-16]. Regulation is a reliability product that corrects short-term changes in electricity use that might affect the stability of the power system. In technical terms, the main goal of regulation is to keep the system's area control error, also called ACE, within acceptable bounds. ACE is the difference between scheduled and actual electrical generation, accounting for variations in the system's frequency. Regulation helps match generation and demand to keep the grid functioning normally by, maintaining a system frequency, tracking moment-to-moment fluctuations in customer electricity use, correcting for unintended fluctuations in the generation (such as a large generating unit disconnecting from the system), and managing differences between forecasted or scheduled power flow and actual power flow on the system.

After restructuring and deregulation of the electricity industry, it is stated that the power system will be more efficient if the differences between peak and low load periods are kept as small as possible [17-29]. It has been demonstrated that the perfect balance between supply and demand in real-time is necessary for the reliable operation of the electricity system. Demand response program (DRP) is defined as changes in electric consumption patterns of end-user clients in response to changes in electricity price over time or to incentive payments designed to decrease high electricity usage at high wholesale market prices times or when the system reliability problems occur. Utilizing the DR concept that not all loads are equally urgent and so less urgent/flexible EV loads may be shifted to other timeslots in a staggered manner alleviating stress on the system. Certain flexible loads, designated as a demand-side response, can be removed from the grid to provide reserves [15, 30]. Fig.2.1 illustrates the role of DR in energy balancing for a typical power system.

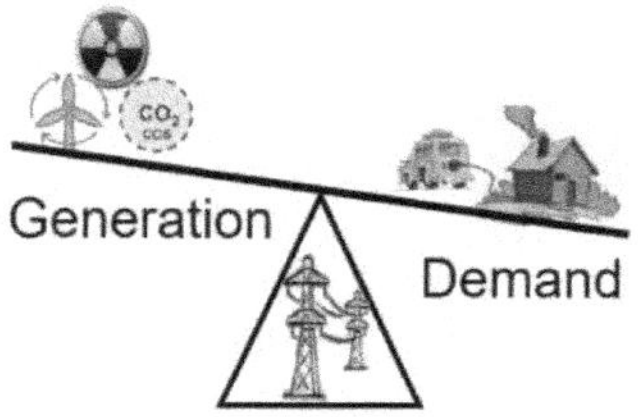

Fig. 2.1: Role of DR in Energy Balancing for a Typical Power System [23]

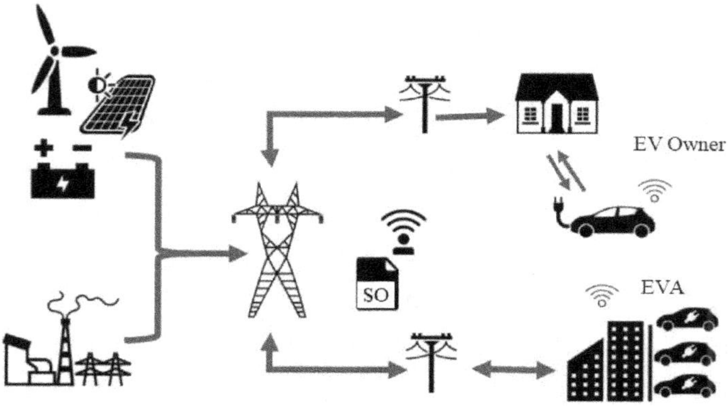

Fig. 2.2: Schematic of V2G Power Flow in the Smart Grid [25]

EVs are idle most of the time in the parking place and therefore can earn revenue by participating in flexibility services and DR programs. Due to their limited size and relatively low impact on the system, the participation of EV owners in DR programs is generally achieved through EVAs [31-47]. EVs can be effectively integrated with the power grid through V2G [10-11, 30, 50-55]. The schematic of V2G power flow in the smart grid is represented in Fig.2.2. V2G has been proven to reduce the EV owner cost, support the power grid, and generate revenues for the EV owner. Due to regulatory and physical considerations, EVAs are necessary for EVs to participate in electricity markets. Restructured electricity markets make possibilities for the EVA to combine the capacities of many EVs and bid their aggregated capacity into energy and ancillary markets. Through market participation, EVA procures electricity in the wholesale market and earn revenues through the provision of reserve and regulation services [33-35,56-57]. It being a commercial entity aims at profit maximization through market participation.

The EV charging infrastructure is formed by a combination of slow AC private (at home) and semi-public (i.e. at shopping malls or office facilities) stations ($P < 7$kW), quick AC public (at the streets) and semi-public stations (kW $7 < P < 44$ kW) and fast public DC stations ($P > 44$ kW). Price incentives might motivate customers to charge their EV off demand peak hours, to enable the mass roll-out of EV and their most optimal utilization.

Table 2.1: Charging technologies [19]

•Level 1 –Typically 1.4kW / 12Amp / 120V
–SAE J1772 connector
–Used in home and workplace charging
•Level 2–Typically 7.2kW / 30 Amp / 240V
–Up to 19.2kW / 80 Amp
–SAE J1772 connector
–Used in home, workplace and public charging
•DCFC–Ranges 36kW / 75Amp / 480V (3-phase AC input) to 350 kW. Typically, 50kW.
–SAE CCS, CHAdeMO, Tesla Supercharger (120kW)
–Used in public, workplace and fleet charging
–Most LDVs currently cannot accept a charge >240kW, although buses may go up to 350kW.

Various charging technologies are illustrated in Table 2.1. Operating at a high charging/discharging rate also harms the battery state of health (SOH). The effects of this factor greatly depend on the EV plug maximum power and the battery capacity: for instance, considering a slow charging station (at home) of 3-5 kW, and a fully electric EV battery capacity of at least 16kWh, the maximum charging rate will be around 0.3C which is pretty low. Batteries would be operating less around extreme SOC values, a slow charging rate is beneficial for the battery SOH. EV charging load by an hour of the typical weekday based on charging location and levels of charging are depicted in Fig. 2.3.

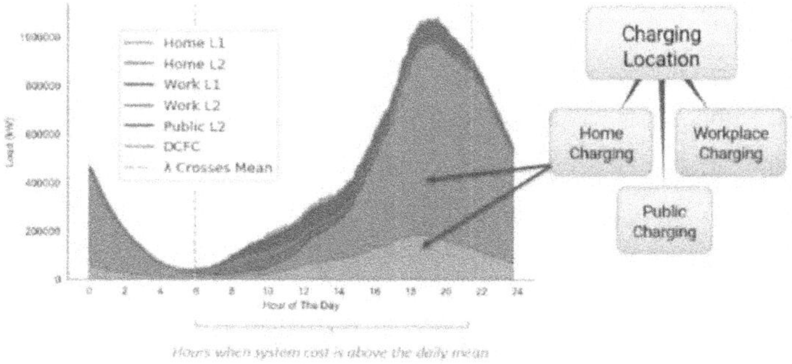

Fig. 2.3: EV Load by an Hour of the Typical Day [19]

EVA has the potential to utilize the flexibility of EVs to serve a range of applications as illustrated in Fig.2.4. United States federal agency and state policy directives Federal Energy Regulatory Commission (FERC) 755 and 841 have spurred EVs mobile storage adoption. FERC 841 opens equitable EVA access to all markets, considering it both as generator and load. The capital cost can be recovered through capacity procurement while operating cost can be recovered through performance/opportunity-based services as illustrated in Fig. 2.5.

A sophisticated strategy is required for EVs mobile storage value maximization through multi-market participation (energy, regulation, and reserve markets). Hybrid models like integrated DR and V2G scheduling could be a possible alternate cost-effective solution to resolve large scale EV and renewable integration intermittency issues. Over time-horizon, EVA can maximize the expected operating income from EVs asset under forecasts of market prices, renewable production and ancillary service performance requirements; while maintaining acceptable asset state-of-health and complying with market regulations, interconnection requirements and tax incentive requirements. It can opt dispatch strategy considering system constraints, PV and EV limits and energy production profiles. G2V charging can occur from Grid-Day ahead energy, real-time energy, Solar PV-clipped energy, economic or technical curtailment, energy meant for pass-through. As battery cycling limitations are imposed, the system priorities cycling only when it is increasingly valuable to do so. The Economics of the V2G business model depends on three things: rapid cost decline, fair access and compensation mechanisms. Sophisticated controls for bundling services and realizing benefits: For example, PV plus EV has strong financial points due to the appeal of free energy (clipping/curtailment) and tax incentives.

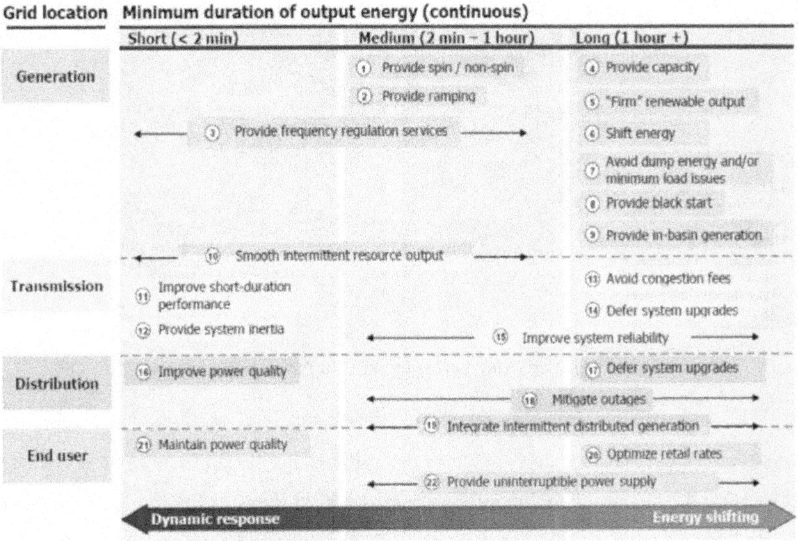

Source: Southern California Edison (SCE)

Fig. 2.4: EVA as a Service Provider at Various Levels and Timescales

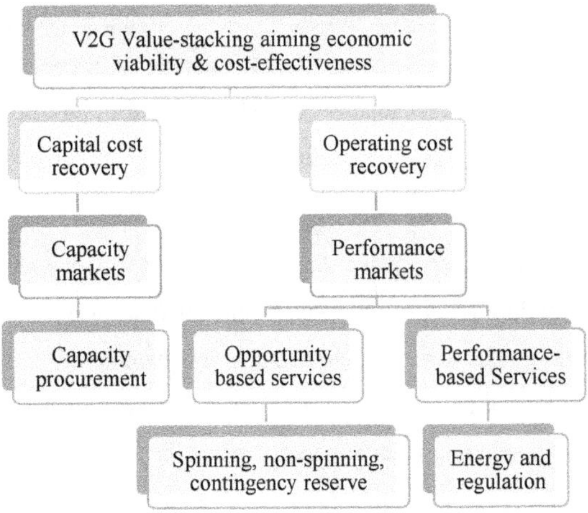

Fig. 2.5: V2G Value-Stacking for Economic Viability and Cost-Effectiveness

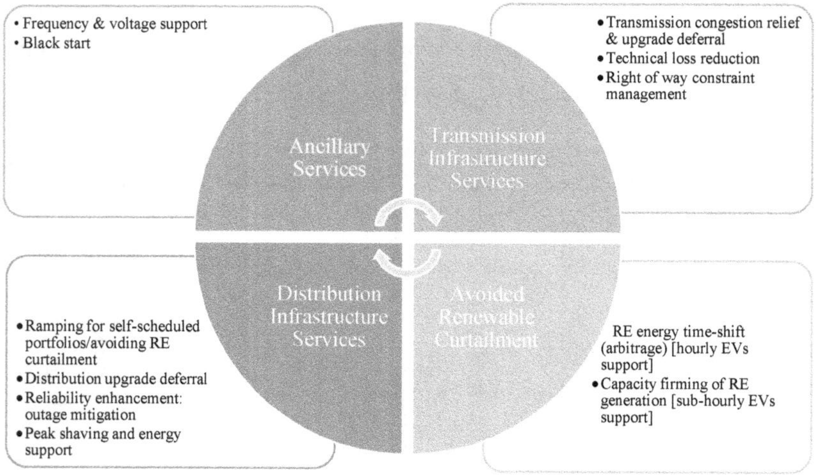

Fig. 2.6: Value Levers for V2G in Power Systems

V2G operational strategies have to balance between maximizing income and minimizing the risks. Fig. 2.6 illustrates the value levers for V2G in power systems to investigate the techno-economic effectiveness of V2G in providing dynamic frequency regulation and other ancillary services. Fig.2.7 depicts the economic value levers of EVA which could be monetized. EVs can be used for ramping support when the RE (solar resources) generation reduces during the evening time. Ramping support by EVA in supporting a rapid decline in RE/uptake in demand.

EVA will enable energy arbitrage by G2V charging when the energy cost is low and V2G dispatch during peak hours. Benefits accrued from energy arbitrage i.e. help in peak reduction by performing charging when the energy cost is low and dispatch during peak (high cost). V2G scheduling is used for deferring distribution capacity enhancements.

Market redesign aspects are discussed in FERC order, section 113 that is relevant to V2G for finding ways to integrate DR into ancillary services markets [23]. The rules of the spinning reserve market are changed by enabling the full participation of DR as a resource. Policy amendments allow demand resources from being full and equal participants in spinning reserve and other ancillary services markets. This is relevant because an EVA is providing regulation is acting as a controllable load (or demand resource) approximately half of the time. The FERC order makes it very clear that demand resources that can be modulated according to an Automatic Generation Control

(AGC) signal from the SO has just as much right to participate in the regulation market as do generators. EVA provides a single, large, stable and reliable power source. Therefore, SO does not see details of single cars, only sees aggregated. EVA schedules capacity in the market and dispatches dispatch requests to EVs. It reports actual power dispatched to SO.

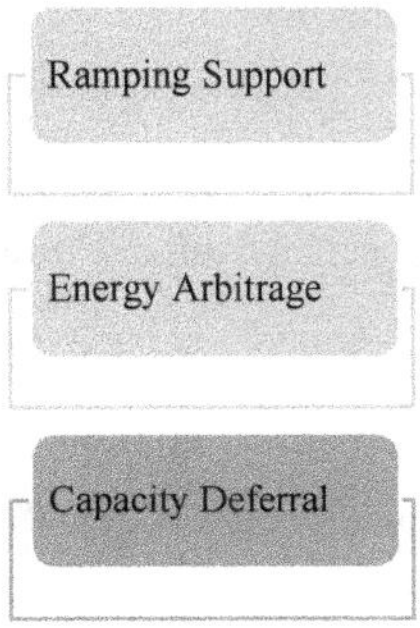

Fig. 2.7: Benefit Streams for EVA

Table 2.2 illustrates the types of EVs and their specifications available and sold in markets. By selecting SO's G2V Regulation Energy Management (REM) network, EVA as a scheduling coordinator, will allow the SO to maintain the EV's preferred operating point (POP) by balancing the energy dispatched from it [11,28,31,35]. Each EV operates at its POP from what it provides upward and downward regulation. The capacity of the regulation services is only linked to the deviation capacity from the POP. However, this approach requires that the EVs only provide regulation in intervals when the SO's dispatching call can accommodate them. POP is the actual power drawn after providing regulation services.

Table 2.2: Types and Specifications of EVs [58]

Vehicle make and model	Vehicle type	Battery capacity size (kWh)	EV range (km)	Max. onboard charge/discharge power (kW)
BMW i3	EV	18.8	130	7.4
Chevrolet Volt	HEV	14	85	3.3
Ford Focus	EV	23	120	6.6
Nissan Leaf	EV	30	172	6.6
Tesla Model S	EV	85	438	19.2

Regulation-up and down dispatch signal should average zero over each charging sub-period since an energy guarantee is to be enforced in unidirectional G2V charge scheduling [28]. The dispatch algorithm, which aims at dispatching power flows among the several aggregated EVs. These aggregated EVs in a coalition is registered as a resource and fully compliant with ancillary service market rules. These EVs take part in the regulation market with an uncertainty of regulation services deployment. In the proposed work, it is assumed that the scheduling algorithm is fully efficient, and therefore dispatch algorithm is not simulated. In other words, the EVA tenders are always accurate regarding the actual available power for regulation, and the SO always accepts the offered price.

In markets such as the Electric Reliability Council of Texas (ERCOT) market and California Independent System Operator (CAISO), regulation up and down services are procured as separate ancillary services by competitive bidding on day-ahead and hour-ahead markets [35,52,59]. Whereas in New York Independent System Operator (NYISO) and Pennsylvania, Jersey, Maryland Power Pool (PJM market) two different types of automated signals are generated to be followed by Regulation Market resources [29]. The Regulation D signal is a fast, dynamic signal that requires resources to respond almost instantaneously. Regulation A is a slower signal that is meant to recover larger, longer fluctuations in system conditions. These two signals communicate with each other and work together to match the system need for regulation. In other words, there are not separate regulation up/down services but combined regulation services in NYISO and PJM markets.

The custom of the EVA's management competencies can be used to partake in "new emergent markets (e.g. "flexibility" markets)" as distributed energy resource (DER) aggregators, in wholesale electricity markets [36, 60]. As stated by the European project Mobile Energy Resources in Grids of Electricity (MERGE), EVA, [37] is an entity that clusters and accomplishes the charging requests of EV owners, and allows their markets participation. The EV resources might vary from restricted EVs in a parking lot to large EVs dispersed in several areas [38]. The role of EVA is being reflected like a path to incentivize EVs adoption. For example, in Spain as per the regulating law in the electricity area (Law No. 54/1997), an EVA is referred to as the system's load manager/load aggregator [18]. It has the potential to buy electricity from the market and sell it to the EV owners. The EVA manages EV charging to follow demand-side management programs set by the government or SO's necessities [39]. The proprietorship of the EVA is visible in the literature to a diverse electricity industry or other segment stakeholders [40]. The EVA could be a new market participant operational as a separate entity or be a share of the present entity in the electricity segment [41]. In the UK EVA exists as a balancing services provider to SO such as Frequency Response, Fast Reserve or Short-Term Operating Reserve [42]. It is a commercial aggregation service provider aggregating smaller loads from many places [43, 44]. To aggregate EV loads/resources and enable their market participation and the provision of ancillary services [45-46], the control approaches that have been proposed for EV management are i) Centralized and ii) Decentralized [61].

The first part provides a comprehensive review of recent research works in the area of V2G scheduling [62-71] and proposes the pragmatic model emphasizing EV owners' perspective. The second part studies the techno-economic issues of EVs in the electricity market and possible management techniques for addressing them [72-76]. The third part conducts application-oriented research to depict the opportunities from grid integrated EVs as a service [77]. The fourth part elucidates the concept and business models for G2V/V2G scheduling of EVA. Further, a comparative analysis is performed for static and dynamic G2V scheduling models. Besides, the necessity for integration of PBDR is defended for the substantial motivation of EV owners for active participation and enhancement of performance efficiency of EVA. Besides, the stochasticity of model parameters is dealt with employing the uncertainty framework of EVA in market participation as discussed in the subsequent section. Simultaneously optimizing perspectives of all involved entities in G2V scheduling poses severe risks to EVA, and requires risk-based G2V decision-making of EVA for multi-market participation. Finally, research challenges and objectives are framed through in-depth discussions and are provided at the end of the chapter.

2.2 Economic & Technical Issues of EVs in the Electricity Market

Techno-economic issues of EVs need to be addressed by EVA for G2V value assessment in the electricity market. EVs have a storage capacity as both a threat to today's SO and an opportunity as if not utilized properly, it will drive up peak demand and may cause grid instability [6-7, 9, 31, 36, 115, 23]. Whereas if used wisely, these EVs can become a distributed energy storage device for SO and hence can be dispatched to lower peak demand, delay the construction of new power plants through intelligent

G2V/ V2G scheduling of EVs by shifting and controlling load demand caused by EVs [30, 44]. V2G facilitates easy integration of RES by scheduling the charging/discharging process to coincide with periods of strong wind or sun [45]. PBDR and flexibility measures could resolve techno-economic issues of G2V scheduling of EVA cost-effectively as shown in Fig. 2.8 [11, 28].

Quantum of existing literature of EV charge schedule is based on adopted RTP- DR [25, 72, 74-76]. Financial rewards of participating in RTP-DR are determined by EV owners' ability to respond to variable prices. RTP-DR mostly in literature has a low acceptance rate by EV owners for being too dynamic to respond and being infeasible for few charge cycles of EVs in a day. It is observed that under RTP-DR, the redistribution of social welfare is not fair as the EVA absorbs the entire benefit due to passing on RTP variations to the consumer [78-79].

TOU pricing scheme is another PBDR program. TOU tariffs are defined by different tariffs for different blocks of hours for a 24-hour day. These blocks vary from two to six in number [80]. TOU tariffs vary depending on the time of day (e.g., peak versus off-peak period) and the season, the weekday and the weekend. They are typically pre-determined by the utility for several months or years. TOU, being less dynamic, is a natural price signal for EVs under DR program. As far as EVA operation is considered, RTP-DR minimises its net payments in the day-ahead and real-time markets. However, to effectively motivate consumers' participation in DR programmes for load levelling, TOU-PBDR is a promising pricing scheme from the customer's perspective [79].

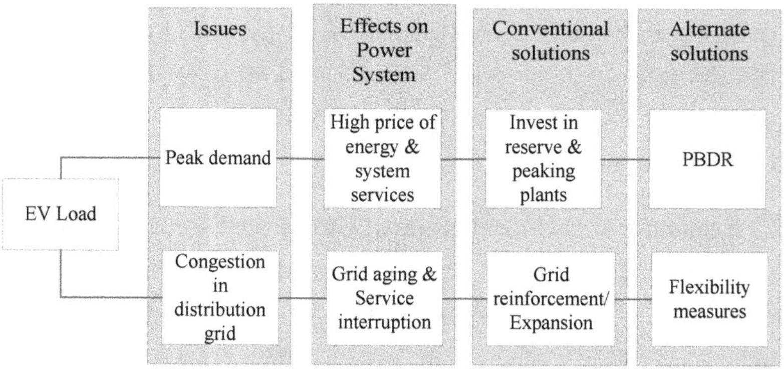

Fig. 2.8: Techno-Economic Issues of Grid Integrated EVs (GIVs) Load and Market-Based Cost-Effective Solutions

TOU-PBDR has the potential to enhance the economic efficiency of the system through the enhanced flexibility of EVs and incentivize EV owners. Charge Scheduling of EVs by EVA may be modelled to maximize EVA's profit while motivating consumers for load levelling through TOU-PBDR at decreased EV charging cost as well. TOU rates tend to mitigate peak load growth by shifting charging load to low-cost hours; however simultaneous charging of EVs during off-peak hours under a TOU schedule may result in a second peak in the load demand if EV adoption is high. To address this, smart coordinated and distributed charging of EV is required [7]. Such charging necessitates EVA not to use a fixed charging rate, rather deviate EV charging rate to distribute the charging of EVs during off-peak times to flatten the daily load profile. The economists believe in the potential of TOU-PBDR to provide economic benefits with customer value enhancement, thus improving the economic efficiency of the electricity market.

TOU, being less dynamic, is the natural price signal for EVs under PBDR but may reduce EVA's profit. It necessitates designing special TOU rates for charging EVs rather than utility TOU, based on charging load profile and pricing information to EVA from EV owners and SO respectively. Considering this, an integrated TOU-PBDR model for EVA's G2V charge scheduling is proposed for the energy and regulation market. EV owner perspective for charging cost minimization is emphasized along with EVA profit maximization. Buying/procurement operation is at RTP while selling to EV owners is at TOU rates specially designed by EVA.

2.3 Opportunities from Grid Integrated EVs (GIVs) as a Service

An optimal and regulated dynamic charging control scheme could co-optimize technical and economic objectives. The area of the EV charge scheduling model has been addressed predominantly in the context of smart distribution systems. However, addressing issues of SO considering EV owners' perspective under dynamic scenario is disregarded. To address this, it requires EVs' full integration in electricity markets and by recognizing the various services that they can provide [77]. Thus, multiple services provision-based G2V scheduling framework for EVA in multiple market segments, could realize EVs' full economic potential and maximize the overall revenue.

EVA achieves valuable charging flexibility from EVs through EV owner-friendly coordinated smart G2V/V2G scheduling. Applying these strategies, EVA can adopt coordinated and distributed charging rates for EVs over the connection hours providing stability, regulation to SO and drive the price reduction attributable to the augmented competitions [31], [27], [50]. Charging flexibility through PBDR is more efficient due to voluntary acceptance and participation. Attractive TOU tariff design can be performed by EVA to avoid rebound effect and/or switching to other EVA.

Integration of PBDR into G2V scheduling can assist in load levelling by providing financial incentives to EV owners to charge at low load levels (off-peak) as compared to the peak load period [27]. This load shifting also manages congestion, increases transformer life and reduces the use of costly plants during peak hours, thereby enhances grid economic efficiency [50, 108]. DR provides the flexibility of selecting viable charging time according to the available plug-in duration and initial SOC level to EVA. However, the full realization of DR potential necessitates the determination of price dynamics offered to EV owners for their active participation [19]. As the charging load of a large EV fleet is huge, EVA buys power from the wholesale electricity market at RTP and sells to the EV owners based on the pricing scheme in PBDR. As EVA is a commercial entity and aims at revenue maximization, the PBDR pricing scheme should also manage the difference between RTP and sell price. Accordingly, RTP with a mark-up can be used in RTP-DR by EVA. However, against flat rate, PBDR integrated scheduling motivates EVs to charge in off-peak hours and utilize their flexibility for load levelling and peak reduction. PBDR reduces EV charging cost as well.

2.4 G2V & V2G Scheduling of EVA: Concept and Business Models

In a V2G system, EVs can connect to the grid, with electricity flowing from the grid to the vehicle (G2V) or back (V2G) and can help solve energy storage and load management problems. G2V/V2G concept is initiated to explore an innovative business model overcoming techno-economic issues from GIVs [45, 122-123].

This survey focuses on viewing this smart coordination from an algorithmic perspective. In particular, important dominating factors for coordinated V2G scheduling from three different perspectives are studied, in terms of SO oriented, EVA-oriented, and EV owners-oriented smart charging. Firstly, for SO-oriented V2G scheduling, various formulations are summarized for load flattening, frequency regulation, and voltage regulation. Secondly, for EVA-oriented V2G scheduling, the algorithmic approaches are categorized by research works as direct/indirect and centralized/decentralized coordinated control approaches. The hierarchical framework is more beneficial than centralized/decentralized from the social welfare aspect. Thirdly, for EV owner-oriented V2G scheduling, based on a commonly shared objective of reducing charging cost, different formulations are proposed by studied research works in terms of customer satisfaction. Moreover, various uncertainty issues, e.g., regulation demand uncertainty, electricity price volatility, EV mobility behaviour uncertainty, etc., have been discussed according to the three perspectives classified. The challenging issues that are commonly confronted during modelling smart coordination are discussed and outlined some future research topics in this exciting area.

The potential energy of EVs that are not being driven represents a power asset deployed by EVA for providing services to SO as illustrated in Fig. 2.9. Some of these services include the potential for transmission stabilization, local voltage support, microgrid support, and using connected vehicles as a dispersed data acquisition system that could report local grid conditions, including voltage, phase angle, and power quality [21, 124].

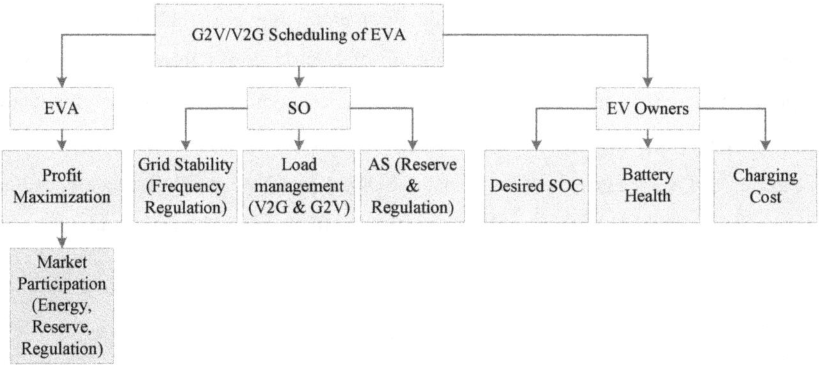

Fig. 2.9: Different Perspectives in G2V/V2G Scheduling of EVA

Geographically categorization of services provided by EVA is done as the local area and wide area services [54-55]. Local area services include peak shaving for demand charge reduction, power quality improvement/ local voltage stabilization, and offering backup power for workplace/homes. The wide-area services provided by EVA involves ancillary services such as regulation and spinning reserves.

A schematic depiction of the V2G system architecture is shown in Fig.2.10. EVA keeps track of EVs that are connected and where they are located with GPS. EVA serves as an interface for each EV owner to permit them to log in and share their constraints and profile information through a web server for charge scheduling. EVA provides ancillary services to the SO and would be paid for that service.

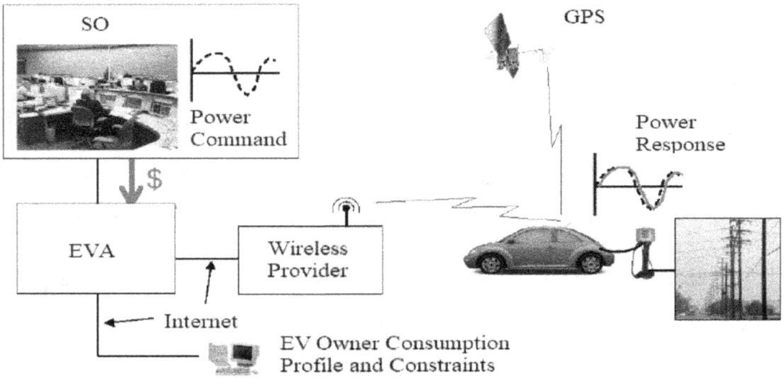

Fig. 2.10: A Schematic of EVA's V2G Architecture for Ancillary Services

Ancillary service prices are paid for two components as illustrated in Fig.2.11, connected capacity only of the available unit, and other for actual energy delivered in real-time. Ancillary services are desired for maintaining grid reliability, balancing supply and demand, supporting transmission of electric power from seller to buyer, regulation service to adjust the grid, specifically local control area to target frequency and voltage [29]. The required amount of regulation service is determined as a percentage of aggregate scheduled demand. SO may purchase ancillary services and/or require individual EVAs to provide an amount commensurate with their loads.

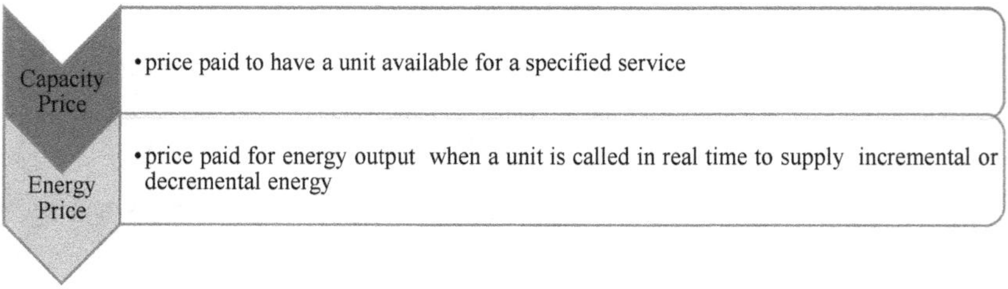

Fig. 2.11: Market Price Constituents for Ancillary Services

V2G may be suitable for peak power in some cases, competitive for spinning reserves, and highly competitive for a regulation [30]. Fig. 2.12 illustrates the various stages of EVA's business models for unlocking the full potential of V2G services. The categories and characteristics of V2G applications are evident in Table 2.3. It depicts the possible power levels, cost, available services, benefits and drawbacks for the unidirectional and bi-directional V2G applications. G2V/V2G has the potential to offer system benefits to SOs and financial benefits to EV owners. Even with its restrictions, G2V is still a rational primary stage in V2G implementation for its trivial cost of charging depots' infrastructure. Lessons learned so far shows that owing to the decreased capital costs unidirectional G2V regulation could be more gainful in certain markets.

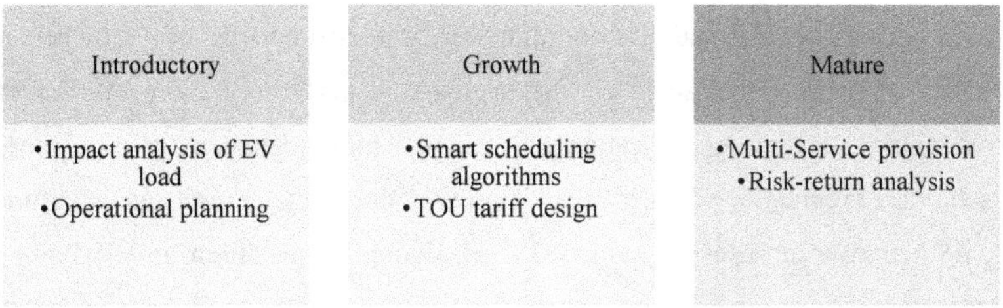

Fig. 2.12: EVA's Business Models for V2G Services at Various Stages

Table 2.3: Categories and Characteristics of V2G Applications

V2G Power Flow	**Unidirectional**	**Bi-directional**
Power Level	Level 1,2 and 3	Expected Level 1 and 2
Cost	Low	High
Available Services	Ancillary Services Classification: • Power Grid Regulation by frequency regulation matching generation & load demand (Direct RT control by SO); Regulation up & down by coordination of SO & EVA; • Spinning Reserve	• Active Power Support • Reactive Power Support • Power Factor Correction • Improve Power System Stability • Harmonic Filter • Frequency Regulation • Energy Back-up
Benefits	• Prevent power grid overloading. • Maximize Profit • Minimize Emission	• Reduce power grid losses • Prevent power grid overloading • Improve load profile • Maintain voltage level • Solution for RE intermittency issue/RE intermittent failure recovery • Maximize Profit • Minimize emission

Drawbacks	• Limited service available "Load only" ancillary services to the grid by varying charging rate.	• Battery Degradation • Complex hardware infrastructure • High investment cost • Social Barriers – Passive participation of EVs for high SOC due to unexpected travelling usage.

2.5 Static Versus Dynamic G2V Scheduling Models

EVA performs load scheduling coordination for mutual benefits of EV owners and SO under a unidirectional static G2V framework. It performs trading in wholesale markets aiming at profit maximization with buy bids for charging energy requirements and sell bids of regulation services to SO. However, due to the unexpected departure of EVs, EVA may incur a loss in static G2V scheduling due to failure in satisfying the charging requirements of EV owners or regulation capacity to SO. It necessitates consideration of a more realistic G2V scheduling method for EVs.

Different parameters considered in V2G scheduling of EVA are classified into 3 groups: Customer-based, market-based, and technical parameters as illustrated in Fig. 2.13. The arrival time of EV is the time when EV is plugged into the charging station of EVA. The departure time of EV is the time when EV is plugged out of the charging station of EVA. The charging period of an EV is the duration in which EV charges/discharges its battery. The initial SOC of EV is defined as the battery energy at the arrival time. The final energy of EV is defined as the battery energy at the departure time. The final SOC is no larger than the battery capacity.

There can be two scenarios of G2V charge scheduling as static/offline/global or dynamic/online/local based on information available to EVA [50, 65-67]. Charging profile information (arrival and departure time, daily energy consumption in trips, EV battery capacity, charging efficiencies, an upper limit of charging rate, initial SOC, trip duration, and distance) is available before static G2V scheduling [68]. However, this information is partially available in the case of the dynamic scheduling model since residual charging energy requirements from initial SOC, and departure times of the EVs that have arrived so far vary due to EV's mobility behaviour [69-70]. Thus, it is the more pragmatic strategy of scheduling. Each EV departure and arrival times vary according to their preferences. Stochasticity in the mobility behaviour of EVs makes the dynamic charging approach challenging to ensure EV owners' driving requirements and operating constraints.

Performance parameters of EVA are sensitive to Round Trip Efficiency (RTE) [169]. EV owners have to bear the costs of energy losses that occur in V2G transactions. $(1-RTE)$ is a performance metric indicating energy losses in V2G transactions or in other words conversion loss as a portion of stored energy lost in each operation of charging or discharging. With the rise in RTE, there is degradation in revenue and cost due to increased losses. RTE Value provides the efficiency of charging or discharging as $Ef_{i_{ch}} = Ef_{i_{dch}} = \sqrt{RTE}$.

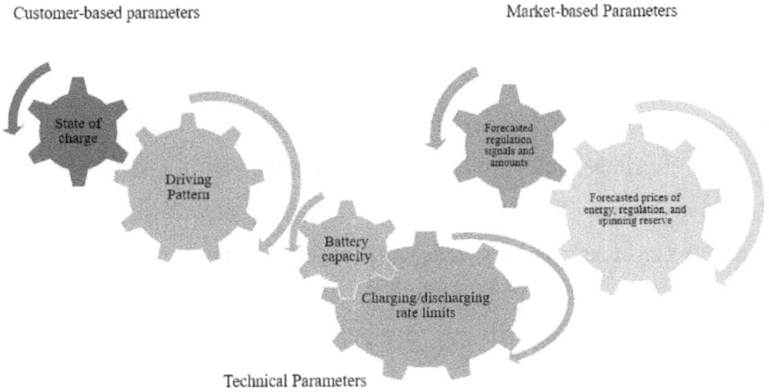

Fig. 2.13: Different Parameters Considered in V2G Scheduling of EVA

2.6 Pricing Mechanism and Demand Response

The historical paradigm for the conventional power System: Demand-driven supply (supply responds to demand). In the smart grid ecosystem, new paradigm/reality: Supply driven demand (demand needs to adjust to meet fluctuating supply with help from storage). Thus, deregulations of the electricity pricing market and advancements in Information and Communication Technology (ICT) have motivated the development of demand-side management (DSM) programs aiming at various objectives from perspectives of different entities as depicted in Fig. 2.14 [17]. Demand Response is a customer action to control load to meet a certain target. Here the customer chooses what load to control and for how long. This is different from Demand Side Management (DSM) where the load is controlled by the electric utility and the customer has no control beyond the initial consent.

EVA can implement more optimal PBDR programs to increase the stability, flexibility, and reliability of the electric grid by managing EV owners' loads. For successful implementation of PBDR, the pricing scheme should be efficient. The pricing schemes are specifically designed for the residential EV owners to actively participate in DR programs to reschedule the charging patterns. Electricity retail prices are designed based on load demand and wholesale market price information. The objective of the pricing mechanism is to minimize the average system cost and rebound peaks. TOU pricing is used to minimize the peak-valley difference, power loss, peak-to-average ratio etc. As aforementioned earlier, uncoordinated charging causes grid instability issues. Price policy should be non-discriminatory and each EV owner should receive a fair electricity tariff rather than a day-ahead retail price, which is based on charging load demand and consumption variation of other EV owners. There are broadly two objectives in designing prices: the primary objective is the problem formulation such that adaptive prices are obtained which are load dependant to achieve even load distribution. It is difficult to implement due to a lack of knowledge of variation in consumer load demand. The secondary objective is to minimize discomfort and energy consumption price by the usage of power in low price hours.

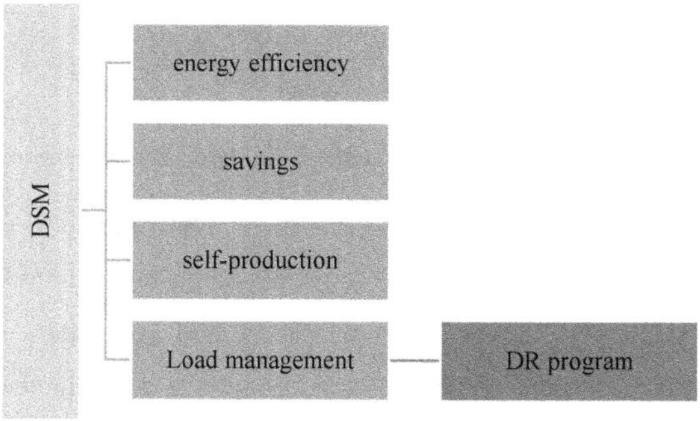

Fig. 2.14: DSM Objectives from Different Perspectives of Market Entities

The focus of existing works is to schedule EV owner load based on market or customized price signal in such a way to minimize cost, rebound peaks and system stability [81]. However, to facilitate EV owners in terms of fair pricing through individualized charging load consumption and pricing policies is still missing and needs to be considered carefully. Fair pricing distribution is capable of considering perspectives of both SO and EV owners, which could trade-off perspectives of SO and EV owners to be feasible for both EVA and EV owners. EVA's revenue should remain unchanged using designed or market prices. Thus, dynamic TOU pricing schemes aim at the motivation of the EV owners to enhance their consumption level during off-peak hours, to improve grid stability through efficient G2V scheduling.

EVA desires to make scheduling smarter, safer, reliable and more cost-effective using advanced distributed computing algorithms, and coordination. The categorization of DR programs is shown in Fig. 2.15 [125]. DRPs are realized by EVA/electric utilities to reduce or shift energy consumption from peak hours of the day when the electricity demand highest to lower demand periods [82]. Moreover, DRPs can help the SO to reduce the price volatility during peak demand hours. Different DRPs can be classified into two main categories: incentive-based programs (IBPs), which are further divided into classical programs and market-based programs, and price-based programs, (PBPs), which are based on the dynamic pricing rates in which the electricity tariffs are not flat. The rates fluctuate following the real-time cost of electricity. The ultimate objective of these programs is to flatten the demand curve by offering a high price during peak periods and lower prices during off-peak periods.

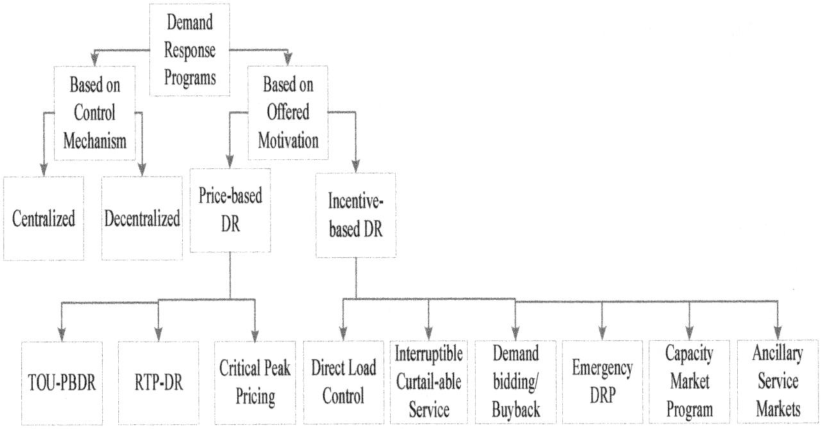

Fig. 2.15: DRPs Classification

In existing researches, the utility TOU prices are usually given in advance, hence without optimizing price structures, the temporal complementarity among EV owners having different load profiles is neglected. Because, without complementarity, it seems difficult to design a flexible pricing tariff depending on diverse power consumption and market pricing trends. Very few researchers have highlighted a pricing mechanism for energy storage units. It necessitates a similar TOU price design model for charging EVs in a smart grid environment.

The price-based demand response programs consist of a time of use, critical peak pricing, and real-time pricing procedures. The basic type of PBP is the TOU rates, which are the rates of electricity price per unit consumption that differ in different blocks of time. The rate during peak periods is higher than the rate during off-peak periods. The simplest TOU rate has two-time blocks: the peak and the off-peak. However, to provide more flexibility to EV owners, the number of blocks must be increased. An optimal number of blocks can be achieved by satisfying the properties of the optimal cluster [83].

The rate design attempts to reflect the average cost of electricity during different periods. In the time of use method, consumers are encouraged to shift a part of their consumption from peak hours to mid-peak and off-peak periods and reduce their bill. The electricity tariffs in the critical peak shifting scheme are higher than the time of use prices. The real-time pricing method is similar to the time of use approach. But hourly electricity prices are considered instead of three-level rates.

2.7 Uncertainty Framework of EVA in Market Participation and Risk-based G2V Scheduling of EVA

EVA faces several kinds of uncertainties as categorized in Fig. 2.16. It faces significant risks due to market price volatility and mobility behaviour dynamics in dynamic G2V scheduling [20, 84-89]. As discussed earlier, ancillary service prices consist of two components: the connected capacity price that is available and the delivered energy price. In the proposed scheme, EVA is paid for available connected regulation capacity. The uncertainty of deployment signals for ancillary service markets is considered when actual delivery of energy to SO is performed under V2G dispatch.

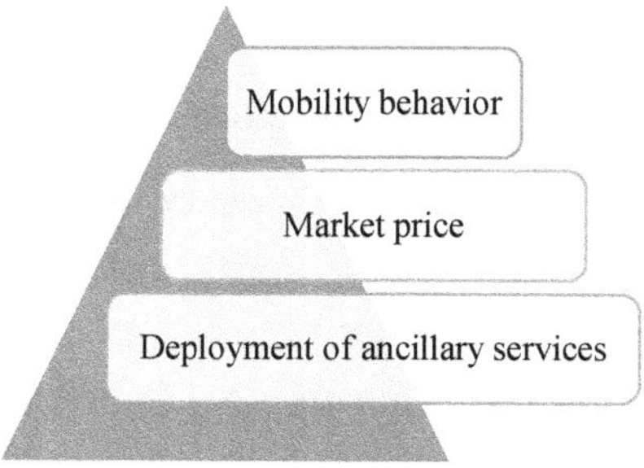

Fig. 2.16: Uncertainties Faced by EVA

Considering perspectives of multiple involved entities emphasizing customer perspective and grid stability, EVA deviates charging rate smoothly from scheduled charging rate (POP). It adopts the risk hedging method of diversification through multi-market participation. Multi-market participation can hedge risks to a certain extent. However, the risk-averse decision-making model of EVA is desired to manage risks from market price volatility and mobility behaviour dynamics. Proper selection of risk measure is required for efficient risk management [68].

2.8 Research Challenges and Objectives

Recent developments in battery technology and smart grid, perceives massive penetration of EVs to the grid, for alleviating concerns on greenhouse gas emissions, energy security, and fossil fuel dependency. Unregulated charging is the baseline scenario of charging whenever and wherever parked, i.e. an EV owner charges its plugged in EV immediately after arrival at home/a charge station. The EV owner keeps charging until its battery pack is fully charged or unplugged for driving purpose. Simultaneous charging scenario, active pragmatically at peak hours, increases the charging cost burden on EV owners. The increased variation of charging rate would lead to the speeding up of battery degradation. Thus, unregulated EVs' charging (concentrated charging along with an increased variety of charging rate) would posture major threats in electricity markets and system/grid operations. It may increase peak loading and necessitates network reinforcement which is not cost-effective. G2V has been intended as a means of lessening the grid potential operational delinquents and enhancing the opportunity for the proliferation of EVs. G2V concept is regarded as a way of swapping ancillary services and energy between the grid and EVs. Therefore, G2V scheduling becomes an emerging research focus with a large-scale application of EVs.

Present market regulations restrict fair participation in power markets for individual EVs' G2V scheduling. EV owners are unable to accomplish the charging regulation for their cars in the most efficient way, EVA as a large load aggregator act as a trading agent between EV owners and SO for intelligent G2V scheduling, subjected to grid and EV constraints as well. Each EV load has parameters lower/upper charging rate limits, amount/time of energy requirements, besides restrictions on the frequency of cycling. Aggregated EV loads will be able to draw some power between these lower and upper charging rate values. The midpoint of this range would typically be bid in the ancillary service market as baseline/nominal/scheduled/expected charging rate/set/ preferred operating point (POP). G2V schedules an EV hourly at the aforementioned optimal charging rate defined as POP which is the targeted average level of power consumption of the EV over a specific period while providing regulation capacities. POP might vary with time and could consider value between zero and maximum charging rate limit. Regulation is a contract over one hour to remotely offer to regulate power up or down from a POP. The amount of regulation contracted is the total amount by which power can deviate from a POP. Regulation up/down dispatch represents reduction/increase of the load. Thus, maximum up/down regulation capacities are dependent on the maximum charging rate limit and POP. Regulation provision from POP modulation of EV charging is based on the regulation requirements from SO [53]. This can be modelled as the probability of call from SO. In the proposed work, this probability is assumed one.

A massive deployment of EVs would necessitate an optimal G2V scheduling framework of EVA to provide flexibleness for electricity markets participation and maximize the benefits of grid integrated EVs. Hourly POP of an EV charging in G2V scheduling affects the perspectives of three entities. A SO's perspective for grid stability through ancillary service provision like up and down-regulation from EV's POP: EVA's perspective of profit maximization and EV owner perspective of charging cost minimization [31]. Optimal scheduling of EVA optimizes POP from its perspective that would influence the benefit of others because their target set-points are disregarded. This necessitates realizing the combined benefits of involved entities (SO and EV owners as well) that requires a co-optimization framework in the G2V charge scheduling problem to enhance the techno-economic efficiency of EVA based on a rational combination of grid and EV constraints [65-66,176]. Quantum work has been proposed on charging control schemes, but most of the approaches focus on maximization of EVA's profit for the energy and regulation market without properly addressing customer requirements [177-179]. Similarly, in other related works, charge scheduling is modelled to minimize charging cost from an EV owner's perspective considering load variance and system losses for SO [159, 180-181]. Considering this, there is a need to enhance the framework for comprehensive coordinated G2V scheduling algorithm of EVA considering maximization of EVA's profit along with simultaneous optimization of monetary benefits to EV owners and provision of regulation services and flexibility benefits to SO.

EVA can adopt coordinated and distributed charging rates for EVs over the connection hours providing stability and regulation to SO. In market-oriented load management, uniform price signals can lead to higher load simultaneity factors and grid congestions. To respond to grid congestion, DR is an effective tool implemented by EVA for improved market operations and allows for load shaping [7]. There is no economic incentive for participation in an incentive-based DR program. PBDR is self-dispatch DR where consumers voluntarily provide load shifting or reductions by responding to time-dependent economic signals, and thus more efficient as compared to incentive-based DR. Integration of price-based DR into G2V scheduling can assist in load levelling by providing financial incentives to EV owners to charge at low load levels (off-peak) as compared to peak load period [7,17,27,144]. This load shifting also manages congestion, increases transformer life and reduces the use of costly plants during peak hours, thereby enhances grid economic efficiency [18-20, 50, 108]. DR provides the flexibility of selecting viable charging time according to the available plug-in duration and initial SOC level to EVA. However, the full realization of DR potential necessitates the determination of price dynamics offered to EV owners for their active participation [19,127].

As the charging load of a large EV fleet is huge, EVA buys power from the wholesale electricity market at RTP and sells to the EV owners based on the pricing scheme in PBDR. As EVA is a commercial entity and aims at revenue maximization, the PBDR pricing scheme should also manage the difference between RTP and sell price. Accordingly, RTP with a mark-up can be used in RTP-PBDR by EVA. However, against flat rate, PBDR integrated scheduling motivates EVs to charge in off-peak hours and utilise their flexibility for load levelling and peak reduction. PBDR reduces EV charging cost as well.

Quantum of existing literature of EV charge schedule is based on adopted RTP- DR [25, 72, 74-76]. Financial rewards of participating in RTP-DR are determined by EV owners' ability to respond to variable prices. RTP-DR mostly in literature has a low acceptance rate by EV owners for being too dynamic to respond and being infeasible for few charge cycles of EVs in a day. It is observed that under RTP-DR, the redistribution of social welfare is not fair as the aggregator absorbs the entire benefit due to passing on of RTP variations to the consumer [79, 91].

The final price of real-time markets are typically known after the actual dispatch period, hence EV owner cannot know the actual price before taking the consumption decision [35, 75, 98, 104, 144]. Hence individually EV owners always take the risk of missing the right signal in RTP-PBDR. This risk to EV owners can be lowered or even completely avoided if they allow an EVA to (optimally) shift or curtail their consumption through another PBDR program as TOU-PBDR. TOU tariffs are defined by different tariffs for different blocks of hours for a 24-hour day. These blocks vary from two to six in number [8]. The TOU rate offerings differ significantly across design features such as the peak-to-off-peak price ratio, number of pricing periods, the timing of those periods, and seasonality. Peak-to-off-peak price differentials of EV TOU rates are generally kept higher than those for a comparable residential TOU rate, for EV owners' acceptance to TOU rates. Depending on the perspectives of the EVA, it may design the rates to encourage charging during low or negatively-priced hours, such as when renewable generation is being curtailed. EVA supporting SO may design rates to discourage charging during specific times when the distribution system is constrained. EVA may view the rate simply as a tool for offering choice and encouraging EV adoption.

TOU tariffs vary depending on the time of day (e.g., peak versus off-peak period) and the season, the weekday and the weekend. They are typically pre-determined by the utility for several months or years. TOU, being less dynamic, is a natural price signal for EVs under DR program. As far as EVA operation is considered, RTP-DR minimises its net payments in the day-ahead and real-time markets. However, to effectively motivate consumers' participation in DR programmes for load levelling, TOU-PBDR is a promising pricing scheme from the customer's perspective [79].

TOU-PBDR has the potential to enhance the economic efficiency of the system through the enhanced flexibility of EVs and incentivize EV owners. Charge Scheduling of EVs by EVA may be modelled to maximize EVA's profit while motivating consumers for load levelling through TOU-PBDR at decreased EV charging cost as well. TOU rates tend to mitigate peak load growth by shifting charging load to low-cost hours; however simultaneous charging of EVs during off-peak hours under a TOU schedule may result in a second peak in the load demand if EV adoption is high. EV specific rate design should consider the pairing of TOU with DR capability; aligning periods of TOU rate blocks with EV load. To address this, smart coordinated and distributed charging of EV is required [7]. Such charging necessitates EVA not to use a fixed charging rate, rather modulate EV charging rate to distribute the charging of EVs during off-peak times to flatten the daily load profile. The economists believe in the potential of TOU-PBDR to provide economic benefits with customer value enhancement, thus improving the economic efficiency of the electricity market.

TOU, being less dynamic, is the natural price signal for EVs under PBDR but may reduce EVA's profit. Literature has shown to use the two-thirds tier TOU prices based on the average of the forecasted RTP and keeping the ratio of peak to off-peak as 2:1 or a trial and ad hoc approach. A poorly structured TOU may result in inadequate exploitation of flexibility benefits arising out of PBDR. Moreover, it may reduce EVA's profit due to inevitable deviations of offered TOU from RTP. It necessitates designing optimal TOU rates for charging EVs rather than utility TOU, considering forecasted RTP and load based on season, weekday, or weekend. Optimal TOU can reflect RTP such that the most significant price variations are captured without comprising too much accuracy in total energy revenue of EVA from customers and save it from avoidable losses. Considering this, an integrated TOU-PBDR model for EVA's G2V charge scheduling is proposed for the energy and regulation market. EV owner perspective for charging cost minimization is emphasized along with EVA profit maximization. Buying/procurement operation is at RTP while selling to EV owners is at TOU rates specially designed by EVA for highlighting customer aspects.

TOU tariff determination depends on (a) setting of a reasonable time interval for different price rates and (b) reflection of RTP appropriately. Some of the literature on TOU pricing investigation [170-171] present the characteristics of TOU tariffs and compare them with RTP to indicate the uncertainty and volatility of prices in RTP. Others give mathematical models to estimate TOU prices [172-175]. The numbers of tariff types and time intervals for pricing are predefined and do not justify the proposed rates settings.

In a static case, when EVA knows charging load and baseload information then it can adopt the TOU design method for the static scenario from [78, 91]. The two key necessities aimed at a viable TOU design are i) to appropriately imitate fluctuations in energy prices and system burdens as well so that EV owners' responses can be suitably directed; ii) to promise the benefits of TOU tariff execution [164-165]. The prices whose period windows are grouped based on energy price volatility and load variation are well-defined as price- adapted and load-adapted TOU tariffs correspondingly. Under such circumstances, the precise rate and design of individual TOU tariff are governed by preserving overall EV owner revenue unaffected related to that under RTP. However, charging demand varies according to different usage pattern for weekdays and weekends, seasons (summer and winter), daily demand duration blocks (on-peak, shoulder-peak, and off-peak). The success of any rate design depends on customer acceptance. The driving forces to shift flexible demands to the off-peak zone are an economic incentive and social conscience [156]. TOU design for PBDR may vary according to the perspectives of SO and EV owners. A performance measure of TOU design can be the satisfaction of EV owners or system security. Mutual optimization of perspectives of both entities gives rise to amalgamated design. However, the amalgamated design of TOU is multifaceted to EVA due to associated acute parameters of zone/tier structure, like length, position, and symmetric/ asymmetric nature. EVA utilizes a clustering technique to design TOU due to its major attributes such as scalability, interpretability, and usability. These attributes are based on similarity metric such as Euclidean-distance, correlation coefficient (centred or non-centred). Clusters are usually groups of nodes densely interconnected but sparsely linked with any other clusters. K-means cluster partitions into pre-defined zones number wise. However, the selection of the abrupt initial mean value of clusters affects severely the results. Therefore, to identify clusters, an efficient and effective agglomerative clustering algorithm based on node similarity is adopted. For a given dataset, first, an agglomerative hierarchical cluster (AHC) tree is constructed based on Ward's minimum variance linkage. Such a tree starts with one cluster and progressively branched out into an increasing number of clusters. This method initially calculates similarities between each pair of nodes, and form pre-partitions according to the principle that each node is in the same cluster as its most similar neighbour. After that, check each partition whether it satisfies the clustering criterion. For the pre-partition who do not satisfy, incorporate them with others that are closest until there are no changes. Therefore, AHCM can better exploit the properties of tariff structure. can obtain superior clustering results compared with other widely used clustering k-means algorithm. Clustering has two decision parameters as a feasible number of tiers and the total number of hours included in each tier [78].

Electricity Price for the tier t_i with 'n_{t_i}' hours (EP_{t_i}) can be obtained from (2.1) as

$$EP_{t_i} = \frac{\sum_{t=1}^{n_{t_i}} Load_t \cdot RTP_t}{\sum_{t=1}^{n_{t_i}} Load_t} \tag{2.1}$$

Where RTP_t indicates RTP for an hour t and $Load_t$ is the power demand by the load for an hour 't' in tier 't_i' with 'n_{t_i}' hours. Granularity index $grain_t$ acts as a performance index in clustering methodology as illustrated in (2.2).

$$grain_t = \frac{\sum_{t=1}^{24}(occurence \quad of \quad one \quad hour \quad clusters)}{24} \tag{2.2}$$

An AHC method has been proposed to determine pricing blocks without pre-knowledge in [78] In this approach, each period in RTP can be clustered into the group in which all the elements have the least price rate difference and, the selected tariff rate number, i.e., the number of clusters in clustering is optimized considering both accuracy and feasibility of implementation. This method has been adopted in this work for TOU design for EVA based on forecasted RTP for a specific season and weekdays and weekend, as per its simplicity and efficiency in the conversion of RTP to TOU.

EVA performs load scheduling coordination for mutual benefits of EV owners and SO under a unidirectional static G2V framework. It performs trading in wholesale markets aiming at profit maximization with buy bids for charging energy requirements and sell bids of regulation services to SO. However, due to the unexpected departure of EVs, EVA may incur a loss in static G2V scheduling due to failure in satisfying the charging requirements of EV owners or regulation capacity to SO. It necessitates consideration of a more realistic G2V scheduling method for EVs.

There can be two scenarios of G2V charge scheduling as static/offline/global or dynamic/online/local based on information available to EVA [50, 65, 67]. The conventional G2V charge scheduling is global/offline/static problem (SCSP). In this method, EVA has prior complete charging profile information from EV owners (times of arrival and departure, initial SOC on arrival and desired SOC on departure) ahead of present timeslot [68]; however, in the dynamic charge scheduling set-up, the charge scheduling plan of EVs is prior not stated to EVA and EV owners are flexible for arrival/departure whenever they desire [69-70]. Consequently, the local/online/dynamic

G2V charge scheduling problem (DCSP) is more realistic in which the EVA does not have EVs information before their actual arrival. The SCSP is regular and simple to model, but the DCSP is more pragmatic and willing and naturally wiser to implement [66]. The charging profile information may be represented as $\left(EV\ ID,\ ta_i,\ td_i,\ SOC_i^{ini},\ SOC_i^{des}\right)$. Each EV departure and arrival times vary according to their preferences. Stochasticity in the mobility behaviour of EVs makes the dynamic charging approach challenging to ensure EV owners' driving requirements and operating constraints.

An optimal and regulated dynamic charging control scheme could co-optimize technical and economic objectives. The area of the EV charge scheduling model has been addressed predominantly in the context of smart distribution systems. However, addressing issues of SO considering EV owners' perspective under dynamic scenario is disregarded. To address this, it requires EVs' full integration in electricity markets and by recognizing the various services that they can provide [77]. Load levelling and active storage capabilities are vital features of EVs and can make EV usage valuable if managed with an intelligent G2V scheduling strategy. Multiple services provision-based G2V scheduling framework for EVA in multiple market segments, could realize EVs' full economic potential and maximize the overall revenue.

With the proper rate structure, EVAs can effectively harness EV charging as a powerful source of DR and load balancing. Additionally, by providing cost-saving opportunities for EV owners, TOU rates can potentially incentivize increased adoption of EVs, thereby facilitating the transition to a decarbonized economy. EV owners are highly responsive to modest price signals and even more so to higher price ratios. The degree of load shifting may increase. TOU rates are highly effective in shifting peak loads, particularly when coupled with smart charging technologies, but energy use and charging load varies widely across different models and charger types. Shifting EV charging to hours in which the grid is less stressed also helps to make better use of the historically underused existing power networks.

EVA acts as a mutually beneficial negotiator between SO & EV owners, enabling improved operations in the power grid and electricity market. It aims to optimally schedule the charging of EVs considering the benefits of SO and EV owner while maximizing its profit earned from participation in multiple markets of electricity and ancillary services. It provides reserve capacity, regulation capacity and grid congestion management through the coordinated charging of EVs for the benefit of SO. It may adopt PBDR to benefit EV owners by reduced charging costs and benefit SO by peak load management and load levelling. TOU PBDR integrated scheduling motivates EVs to charge in off-peak hours and utilise their flexibility for load levelling and peak reduction. TOU-PBDR reduces EV charging cost as well. TOU-PBDR shields EV owners against price volatility. However, for optimal decision making, EVA is subjected to risk from volatility in different market prices.

EVA is also subjected to challenges like its capability to attract and sustain EV owners and to make certain that the combined capacity of idle EVs is available to SO [138]. Multiple market participation hedges EVA against the risk of profit loss as compared to single market participation [77]. However, the incorporation of a risk model based on different uncertainties is necessary for realistic decision making in the G2V scheduling.

The scheduling model of EVA is demonstrated in [139-140] based on minimization of charging energy cost and the net substation cost subjected to network security constraint. DRPs act as a hedge against high peak prices and motivate EV owners to charge during off-peak. In an intelligent grid, DRP plays a dynamic role to re-establish the equilibrium regarding electricity supply and demand [139]. Integrated DRPs and risk-based G2V scheduling are necessitated in the evolving smart grid scenario [101, 141-142]. In [143], G2V scheduling of EVs is modelled using the stochastic approach given uncertainties of the electricity market and EVs to reflect the influences of diverse DRPs on the operative performance of EVAs and to enhance the contribution of EVAs in respective DRPs. The stochastic planning model enables the decision-maker to specify the risk level caused by uncertainties in renewables (wind and solar), load patterns, parking patterns, and transmission lines' reliability, and balance its costs & benefits according to this risk level [85].

The Risk-aversiveness of EVA in market participation should be optimized to attain an enhanced scheduling approach that assurances a Pareto upgrading on its profit over the entire uncertainty set. Measures of risk are necessary for exemplifying the risk related to a given judgment [90, 95]. The most efficient risk measure Conditional Value at Risk ($CVaR$) is a coherent and downside risk measure. Risk-averse bidding has been formulated for demand-side resource aggregator based on $CVaR$, to minimize expected regret value over a subset of worst-case scenarios whose collective probability is no more than a threshold value [96]. Multi-market participation of an EVA is formulated for EVA's $CVaR$ maximization in [87-89]. The risk-averse formulation illustrated that the EVA would be able to reduce the risk in its expected revenue by shifting its revenue source from one market to another market [107]. However, DRPs are not integrated into [87-89, 108-109].

Several methods are used in financial literature and technical engineering for risk management of different market participants such as generation companies, retailers, distribution companies, large consumers, load aggregators such as mean-variance technique, value at risk (VaR) approach, and Conditional Value-at-Risk ($CVaR$) methodology. However, VaR and mean-variance techniques are not suitable in stochastic optimization, as they may lead to multiple extrema (the former) or may have computational intractability (the latter). Although, VaR is a simple convenient representation of risks, as $(1-\alpha)$ -quantile of the profit distribution, a "fat tail" appearing in profit distributions is not detected by the VaR. Moreover, it does not satisfy the subadditivity property of coherency and thereby portfolio diversification may increase the risk. $CVaR$ is another percentile risk measure for downside risk, more conservative than VaR due to consideration of risks beyond VaR. It is an improved version of VaR alleviating the shortcomings of VaR. It is a coherent risk measure and can be calculated easily for any symmetric as well as non-symmetric distributions. The $CVaR$ at a confidence level, $\alpha - CVaR$, is equal to the expected profit in the $(1-\alpha)\times 100\%$ worst scenarios. It provides stable statistical estimates and shows consistency at different confidence levels compared to VaR.

The problem is a mixed-integer non-linear programming (MINLP) problem and is formulated in GAMS [137]. Solver selection is based on its accuracy and computational efficiency. Algorithms based on the Simplex method search vertices of the feasible region; therefore, for large-scale linearized problems, in which the number of vertices is high, the solution time is considerable. As a result, Simplex is categorized among exponential time algorithms. In other words, the time needed by Simplex for solving an LP problem is an exponential function of the problem scale. The interior-point algorithm's search path is inside the feasible region; in fact, central points should be investigated. Thus, the problem scale is not too critical and it can be considered among polynomial-time algorithms. Therefore, COIN-OR IPOPT (Interior Point Optimizer), an open-source solver capable to solve a large-scale MINLP problem is employed to solve the proposed problem.

The grid integration of EVs and associated G2V scheduling of EVA is an effective research area in which numerous works have been published [7, 35, 65-66, 75, 98, 104,144]. [145] considers multi-market participation considering EV owners' perspective of charging cost minimisation using SCSP which is not realistic. [66] models the coordinated DCSP in the energy-only market from SO's perspective of energy loss minimisation considering mobility behaviour uncertainty while ignoring EV owners' perspective. [65] models multi-market participation of EVA through coordinated SCSP and DCSP emphasising on EV owners' perspective. However, it considered RTP-PBDR which is less acceptable to EV owners as compared to TOU-PBDR due to shielding from price volatility risk. Through the implementation of TOU-PBDR, simultaneous optimisation of EV owners and SO can be observed [7, 144]. In [7], a special TOU is designed from SO's perspective of mitigating adverse impacts of charging in terms of additional voltage drops. [144] models valley filling (for SO) and reduced battery degradation (for EV owners) through coordinated TOU-PBDR from utility offered TOU rates. However, [7, 144] modelled SCSP for the energy-only market. [35, 145-146] checks the viability of EVA's deterministic business model multi-market participation using the SCSP approach which is not pragmatic. [143,147] incorporates a stochastic optimisation approach neglecting DR and risk assessment. It was validated that smart charge scheduling approaches can defer network reinforcements, assist through congestion management and reduce energy losses, and alleviate renewable energy curtailments. However, EVA faces multiple uncertainties for G2V scheduling comprising the volatility of electricity market prices and dynamic mobility behaviour of EV owners. Such uncertainties could pose financial risks to EVA's market operations.

On the supply side, EVA faces regulation and energy prices uncertainty. On the load demand side, EVA is compelled to negotiate EV owners' variable charging load demand. Consequently, EVA is subjected to the ambiguities of EV owners' charging load demand. Different volatilities of power market participation yield the call for the execution of risk assessment in G2V scheduling. It necessitates risk assessment in DCSP. Energy only market participation of risk-seeking and risk-averse EVA using

SCSP approach is modelled considering energy price uncertainty while ignoring mobility behaviour uncertainty, DR and EV owners' perspective [108]. [87-88] modelled risk-based behaviour employing SCSP for multi-market participation neglecting DR implementation and EV owners' perspective. [83, 91] discusses the AHCM design methodology of TOU prices for PBDR implementation. Using this AHCM approach, [120-121] models integrated TOU-PBDR and DCSP for multi-market participation of EVA considering benefits of all entities involved (SO, EV owners, and EVA) in G2V scheduling. However, through TOU-PBDR, EVA is subjected to price volatility risk, and therefore it necessitates financial risk management for EVA's DCSP framework integrating TOU-PBDR, with mobility behaviour uncertainty as well as multiple market uncertainties.

The National Household Transport Survey (NHTS) consists of 110,000 interviews conducted in New York over the last 10 years, describing daily travelling. It is found that cars in New York on average are driven for 50 km per day [93]. The overall commute is thus split into two trips per day. The morning commute is at 7:00-9:00, while the return is at 17:00-19:00. The working hours are 8 hours per day which is assumed to be between the period 09:00-17:00. The average distance driven per car is higher on workdays than at the weekend, but the distance for an average day is used without differentiating between workdays and weekends. Measurements were made for the consumption of energy during driving. The result is that the return trip consumes 8.8 kWh. Also, the car is not charged after reaching the workplace but only once it returns to the household.

EVA can facilitate market participation of EVs and provide regulating power, frequency-control reserves, and voltage support services. Most of the research works focus on the impact of EVs integration on the operation of electric distribution networks and battery degradation, and congestion management within electric distribution networks, disregarding customer perspective. Smart coordination among the SO, EVA, and EV owners can bring various benefits to all parties involved, e.g., improved reliability and safety for the SO, increased profits for the EVA, as well as enhanced self-benefit for EV owners. Fig. 2.17 illustrates the research objectives flow diagram.

The first work that we explored in this thesis is the static G2V charge scheduling algorithm of EVA incorporating RTP-DR, which is proposed considering the benefits of all involved entities with complete charging profile information available to EVA in advance. It aims at its profit maximization through participation in energy and ancillary service markets to support grid stability. The problem of charge scheduling of a single EVA is formulated under different EV owner and SO constraints. EV owners' perspective of charging cost minimization is emphasized posing an upper limit on charging cost.

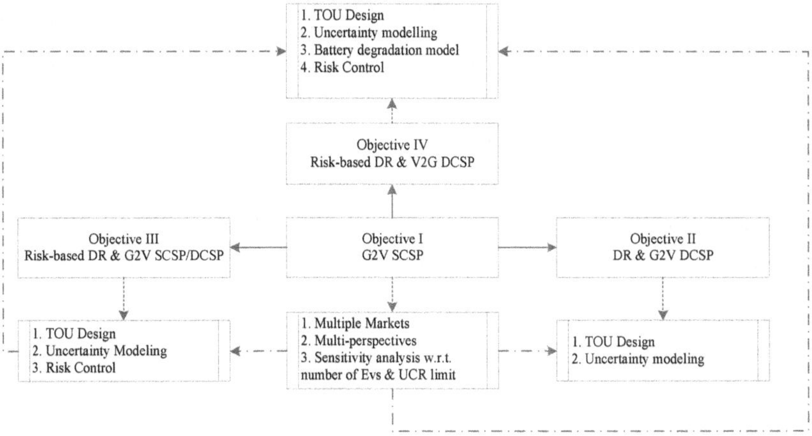

Fig. 2.17: Research Objectives Flow Diagram

Static charge scheduling strategy is not pragmatic due to neglecting mobility behaviour uncertainty. Subsequently, it necessitates the data on the upcoming base loads and the arrival times and the charging periods of the EVs that will arrive in the future time of the day. Nevertheless, the dynamic scheduling strategy accomplishes a slight inferior to the static charge scheduling strategy, it is a real-world scheme that can proficiently handle a large EV population and dynamic EV arrivals. Therefore, the dynamic scheduling strategy is the ultimate determination recommended in this work. Market price risk is passed on to EV owners in RTP-DR. Customers can be shielded from price volatility in TOU-PBDR. TOU being less dynamic is a natural price signal having high acceptance from EV owners. TOU-PBDR can enhance the value of EV owners. Utility TOU-PBDR is acceptable but may cause the creation of another peak called rebound effect due to coincident charging. Therefore, to increase the efficacy of the proposed problem, special TOU prices are designed by EVA using well established AHC method for providing more flexibility to EV owners and satisfying regulatory constraint as well [78]. In this regard, this thesis highlights a more realistic integrated TOU-PBDR and dynamic G2V charge scheduling algorithm of EVA under different

EV owner and SO constraints to optimize EVs coordinated charging and earn monetary benefits for all stakeholders such as EV owners and SO. The effectiveness of the proposed model in terms of peak load management and monetary benefits for all stakeholders is highlighted. Moreover, offering regulation services to SO can support grid stability and thus avoid inevitable contingencies.

EVA may adopt PBDR to benefit EV owners by reduced charging costs and benefit SO by peak load management and load levelling. However, for optimal decision making, EVA faces several uncertainties like volatility in different market prices, EV arrival and departure times, initial SOC based on unexpected trip plans, renewable distributed generation (DGs) output at the charging station, system load etc. It is also subjected to challenges like its capability to attract and sustain EV owners and to make certain that the combined capacity of idle EVs is available to SO [20].

In recent times, the volatilities of the loads and market prices are modelled considering approaches (MCS, time series forecasting, artificial intelligence analysis) [92, 118], MCS approach has received more attention for easy to implement while providing an approximate solution to both stochastic and deterministic mathematical problems [119]. Besides, the use of the MCS mechanism has been promoted due to lesser computational costs, mostly accepted that returns are arbitrary and a requirement to offer more robust fiscal strategies to decision-makers. Therefore, the MCS technique is imitated here to characterize the uncertain parameters of market prices pattern and EV owners' arrival and departure behaviour. The higher number of scenarios could well capture the volatility behaviour leading to a more accurate and pragmatic solution. Nevertheless, analysing the aforementioned whole set leads to the intractable problem along with huge computing times. Considering this, an effective scenario reduction algorithm [126] is necessitated to retain the pattern of uncertainty intact. KD matrix scenario reduction technique is one of the prevalent schemes and is employed in the present work for EV mobility behaviour and market prices scenario reduction [92, 119, 126]. It deletes and aggregates the scenarios by KD matrix, which is the likelihood distance between two different scenario subsets that characterises the comparable stochastic procedure. The flow diagram of the uncertain parameter's scenario generation as well as reduction, and the mathematical steps for scenario reduction can be referred from [120].

Multiple market participation hedges EVA against the risk of profit loss as compared to single market participation [85]. However, incorporation of a risk model based on different uncertainties is necessary for realistic decision making in the G2V scheduling [74]. The scheduling model of EVA is demonstrated in [90, 96] based on minimization of charging energy cost and the net substation cost subjected to network security constraint.

EVA performs coordinated charging taking care of its consumers' comfort level. EVA acts as a buffer between EV owners and the wholesale electricity market, assuming the risk of consumption deviations from schedule. Integration of DR can better achieve the advantage of the EV owners' flexibility by shifting or reducing their power consumption. Furthermore, DR can support a reduction in price volatility by shifting power consumption to periods characterized by low prices and high renewable power production [109]. DRPs act as a hedge against high peak prices and motivate EV owners to charge during off-peak. In an intelligent grid, DRP plays a dynamic role to re-establish the equilibrium regarding electricity supply and demand. PBDR offers EVA the possibility to simultaneously decrease its risk and costs on the market. To make the most of the different types of commodities being traded in the electricity market (energy, regulation, and reserve) to maximize the economic performance of the EVA, thus increasing its profitability and competitiveness.

On the supply side, EVA faces regulation and energy prices uncertainty. On the load demand side, EVA is compelled to negotiate EV owners' variable charging load demand. Consequently, EVA is subjected to the ambiguities of EV owners' charging load demand. Different volatilities of power market participation yield the call for the execution of risk assessment in G2V scheduling. Energy only market participation of risk-seeking and risk-averse EVA using SCSP approach is modelled considering energy price uncertainty while ignoring mobility behaviour uncertainty, DR and EV owners' perspective [108]. [87-88] modelled risk-based behaviour employing SCSP for multi-market participation neglecting DR implementation and EV owners' perspective.

Integrated DRPs and risk-based G2V scheduling are necessitated in the evolving smart grid scenario [87-89]. In [107], G2V scheduling of EVs is modelled using the stochastic approach given uncertainties of the electricity market and EVs to reflect the influences of diverse DRPs on the operative performance of EVAs and to enhance the contribution of EVAs in respective DRPs. The stochastic planning model enables the decision-maker to specify the risk level caused by uncertainties in renewables (wind and solar), load patterns, parking patterns, and transmission lines' reliability, and balance its costs & benefits according to this risk level [108]. Uncertainties associated with energy and regulation prices, and the behaviour of EV owners based on their driving patterns, are considered in the proposed stochastic method and validated through several case studies.

The method of realizing anticipated results as per the preferences of decision-making EVA, by decreasing the likelihood of failure/catastrophe with consideration of uncertainty is termed risk management [126]. Uncertainty deliberations offer informed decision support for future market participation planning of EVA. Unrestricted risks may result in significant destruction; henceforth it becomes necessary to perform rational scheduling to manage risk for acquiring the profit maximization of EVA. An EVA ready to yield prospects arisen from uncertainty may want to take higher risk to maximize its profits. On the other hand, it may take a rational decision to limit its exposure to risk at the cost of lower benefits.

Risk management suggests a range of strategies for different desired balances between risks and returns [111-117]. Risk management is based on the appropriate measurement of risk. Several methods are used in financial literature and technical engineering for risk management of different power market partakers such as generating plant owners, retailers, distribution companies, large buyers, load aggregators, for instance, modern portfolio theory, value at risk (VaR) approach, and Conditional Value-at-Risk ($CVaR$) methodology. However, VaR and modern portfolio theory is not suitable in the random process optimisation, as they might draw into multiple extrema (the earlier) or might have computational intractability (the subsequent). $CVaR$ is one more percentile risk quantifier for downside risk, more conservative than VaR due to consideration of risks beyond VaR. It is an improved version of VaR alleviating the shortcomings VaR. It is a coherent risk quantifier and could be calculated efficiently considering all symmetric as well as non-symmetric distributions. The $CVaR$ about confidence interval, $\alpha - CVaR$, is corresponding to expected profit in $(1-\alpha)\times 100\%$ the worst scenarios. It provides stable statistical estimates and shows consistency at different confidence levels compared to VaR.

[83, 91] discusses the AHCM design methodology of TOU prices for PBDR implementation. Using this AHCM approach, [120-121] models integrated TOU-PBDR and DCSP for multi-market participation of EVA considering benefits of all entities involved (SO, EV owners, and EVA) in G2V scheduling. However, through TOU-PBDR, EVA is subjected to price volatility risk, and therefore risk assessment is desirable in the decision making of DCSP. It necessitates risk assessment in DCSP. Therefore, the problem of charge scheduling of EVA under different uncertain parameters has been formulated and designed risk-averse coordinated charge scheduling algorithms for EVA to obtain a trade-off between profit and risk. Risk management is executed to obtain trade-off solutions between profit and risk. In this chapter, a stochastic optimization model is presented for risk-averse integrated TOU-PBDR and dynamic G2V charge scheduling of EVA in the day-ahead multiple markets viz. energy and regulation service provision. EVA participates in these multi-market aiming at profit maximisation through dynamic charge scheduling. The times of arrival and departure of EVs and hourly market prices are considered as potential uncertainties, which are represented in MCS scenarios. The *CVaR* is utilised as a risk quantifier for managing EVA's financial risks caused by the uncertainties. Risk-averse EVA's dynamic G2V scheduling strategy is formulated in General Algebraic Modeling System (GAMS) modelling tool. Accordingly, EVA maximizes its expected profit and *CVaR* in the risk-averse decision-making problem while minimising the charging cost from EV owners' perspective.

V2G can be considered as an extension of controlled charging and one of the potential approaches to handle the issues from the uncontrolled operation of EVs [45, 150]. As mentioned earlier in this strategy, bidirectional power flow exists between the EV and the power grid. The role of an EVA coordinating the charging/discharging of EVs in V2G scheduling comprises of two aspects: 1) To incentivize EV owners for providing V2G services, which represents an economic perspective, and 2) To provide effective DRPs to SO, which represents a technical aspect. An amalgamation of V2G operation and PBDRP can be designed considering centralized, decentralized, and hierarchical control strategy [151]. Hierarchical V2G scheduling of EVA is more realistic due to the capability of efficient control even under incomplete charging profile information and large-scale integration [8, 61, 152-153].

Since PBDR plays a key role in optimum asset utilization and to defer the need for new infrastructure investment [154]. Besides, V2G scheduling integrated with PBDR may avoid inflation of demand peaks by utilizing monetary incentives and price-responsive behaviour of EVs [155]. Furthermore, PBDRPs can reduce the risk of market participation, increase the profit of flexible players, and improves the reliability & efficiency of the system as well [156]. However, it raises the following research questions towards EVA:

RQ1: How to achieve increased battery life in V2G operations with its exhaustive use?

It can be answered through the modulation of the charging/discharging rate to stagger V2G operations over plug-in duration is required to improve the life of an EV battery. Also, EVA could exploit V2G with EV load controllability, storage capability for efficient market participation, and strengthen grid reliability as well [157].

RQ2: How much the economic incentives to be provided to EV owners in TOU price based V2G scheduling to motivate EV owners to participate actively in DRP?

The lack of knowledge among EV owners on how to respond to time-varying prices may considerably reduce their participation in DSM programs. By offering attractive prices to EV owners for incentivizing them and to boost the willingness to participate in DRPs. Various prices have been offered to EV owners for incentivizing them to boost the willingness to participate in DRPs. RTP/CPP adoption in DSM programs to adaptively control the electricity consumption face challenges such as computational complexity and the need for centralized control for simultaneous optimization of benefits of EV owners and SO. In TOU, the energy price is varied during different hours of the day. However, since the price during each time interval is independent of the real-time behaviour of EV owners and multiple EVs may charge during off-peak hours coincidently. This can cause an undesirable rebound peak/avalanche effect [158].

EVs are likely to be distributed ES and flexible PBDR resources for providing ancillary services to SO through EVA. As an interface between SO and EV owners, EVA felicitates smart coordination over operational V2G schedule in multiple markets for reliable regulation provision. Execution and transaction of integrated PBDR and V2G raise economic threats to V2G coordinator/an EVA in case of utility designed TOU prices. To resolve this problem, EVA designs TOU rates based on the forecasted market prices. Optimally designed TOU price provides incentives to customers and protects them from uncertain market prices. Specially designed TOU PBDR is integrated for improved market operations, to reduce charging costs of EV owners and load levelling via enhanced flexibility to SO. EVA manages the charging-discharging scheduled to participate with limited energy in feasible markets. EVA assists for grid stability & reliability through ancillary services, varying charging/discharging rates according to system security constraints.

As per the guidelines and standards for the development of EV charging in India from the Ministry of Power [201], the tariff for supply of electricity to EV charging at a parking lot will not be more than the average cost of supply plus 15%. Therefore, to make the most out of aggregated EVs, EVA can utilize their charging flexibility and operate EVs in G2V/V2G mode at the off-peak/peak hours of energy price. V2G for power markets may be suitable for peak power in some cases; competitive for spinning reserves; highly competitive for a regulation [30, 150]. Therefore, optimal V2G scheduling of an EVA for market participation has been developed in various directions according to perspectives of involved entities SO and EV owners and has recently become a focal point of interest. However, as discussed in G2V scheduling, the dynamic mobility behaviour and market price volatilities pose challenges to the V2G scheduling of EVA for its profit maximization, making it a more complex problem of decision making under uncertainty from a computational perspective. The art of dealing with uncertainties and associated modelling techniques for the decision-making process are introduced and compared along with demonstrating their strengths and weaknesses [88, 182].

Quantum of work concerning analyzing and modelling the problem of EVA's V2G scheduling considering different uncertainties has been done in recent years. The optimal charging profiles for the power grid considering the risks of wind and solar variability [183]. In [184] a stochastic bidding framework for a V2G-enabled operator participating in day-ahead markets is formulated to schedule the operation of EV fleets against various sources of uncertainties, e.g., the arrival and departure time of PEVs, solar power and real-time prices while maximizing the net profits from day-ahead markets. The coordination of the V2G-enabled EVs can improve the operator's revenues owing to the load levelling and regulation services in multiple markets. However, risk management disregarded.

The risk-constrained stochastic optimization-based decision framework for EVAs participating in day-ahead multiple markets viz. energy and regulation markets is modelled leveraging $CVaR$, minimizing the operation cost of the EVs while satisfying the market constraints and EV charging requirements [185]. EVA's perspective is to willingly provide energy and regulation services to the market during peak hours and buy energy from the market during the low-price hours. The risk-aversion/weighting parameter defines the trade-off between the expected cost and the risk of cost variability (i.e., $CVaR$). However, the impacts of EVs on the network operational limitations in the distribution system are not analyzed. Also, the DR framework enhancing customer perspective is not modelled. Under the smart city planning concept, the V2G framework makes it possible location-wise charging profiles of EV owners like workplace charging, recreation place charging, night charging at home etc. [186]. With the help of V2G technology, EVs can be aggregated to behave as a storage system, yielding both economic and environmental benefits. However, several challenges and barriers related to the use of EVs for grid ancillary services need to be analyzed and solved urgently, as the number of EVs is increasing sharply.

Large scale EVs stresses highly a power grid, especially when the supply cannot cover and actively respond to the unpredictable demand caused by charging EVs. Ancillary services of EVs have led to the idea of utilizing EV batteries for grid support, owing to their high-speed response to the fluctuating power system. In [187], a techno-economic analysis of the V2G system with PBDR is conducted in terms of the changes in the tariff schemes. EVs owned by business entities as operating vehicles with the dynamic tariff is feasible for ancillary services, and can potentially reduce the cost of charging. From a power company perspective, V2G also potentially improves annual revenue. However, it neglected risk management due to price volatility. [188] reviewed the interactions of EVs with power grids through coordinated risk management of EV charging infrastructures. It identifies and reveals the potential for mitigating initial investment costs, beneficial to develop reliable power demand prediction models and scheduling techniques to maximize V2G benefits. Scenario-based analysis and risk management may assist in infrastructure development until economic factors make EVs a preferred option. DR implementation from EV owners' aspect is ignored. [189] determined viability of the business model through financial risks associated with accelerated battery degradation in a V2G-enabled fleet. V2G opportunities may present additional revenue for EV owners provided that the burden of driving and V2G services do not accelerate the degradation of the battery systems in these vehicles. Operation stability and economic performance of microgrid are improved with optimal bidirectional V2G mode of EVs. The economy of the system has been quantified [190]. However, risk management & DR integration overlooked. [191] Look-ahead risk-averse power scheduling of V2G and G2V systems enabled EVs is formulated to model the coordination between EVAs and DISCO in an eco-environmentally friendly manner considering severe uncertainty, where the main objective is to maximize the profit of EVAs constrained by operating costs of the system. It did not perceive the EV owners' aspect. Expert scepticism and consumer distrust of V2G is highly prevalent, necessitating better communicate the benefits of V2G to both entities [192]. Techno-economic analysis of V2G feasibility is investigated through stochastic security-constrained unit commitment [193-194]. Risk-based V2G/G2V scheduling strategies for the risk-averse and risk-seeker EVA are modelled considering mobility behaviour uncertainty and electricity price volatility in the day-ahead market. A risker decision leads to a higher profit. However, the customer perspective ignored. V2G deployment is essential for sustainable development and managing risks of power grid operation. Risk-involved stochastic scheduling of EVA in day-ahead and reserve markets is studied using the downside risk constraints method to consider the risk associated with uncertainties of market prices via scenario approach [195-196]. The energy fairness problem and the need for broadening the cases of V2G systems is particularly emphasized. The risk-averse state of the model is compared to risk-neutral which in the former one the profit is reduced meanwhile that risk-in-profit is declined. However, PBDR is ignored [193-196].

Risk management of smart grids based on the V2G charge scheduling strategy of EVs is executed using Monte Carlo simulation [197]. Impact analysis of charging and penetration level of EVs on distribution system are analysed in terms of uncertain parameters. In the V2G scenario, EVs acted as storage systems supporting the system. EVA is an agent for facilitating the interaction between SO and EV owners, which could bring advantages for all of them. Not only could the EVA participate in the day-ahead market as a representative of EV owners, but the EVA could also manage the integrated load of EVs via arranging the charging and discharging times in responding to price signals. For optimal grid management, a coordinated charging/discharging load model for EV with TOU price mechanism is designed to balance the optimization between system power loss and node voltage fluctuation [198-199]. The technical challenges to V2G and the potential benefits of V2G are discussed [200]. Multi-objective resource scheduling of a smart distribution system including a large number of EVs' charging/discharging management system is proposed to reduce both operation cost and air pollutant emissions. However, risk management and PBDR is ignored.

Quantum work has been proposed on V2G scheduling of EVA but most of these schemes focus on the individual objective without properly addressing the operational behaviour of EVA. Furthermore, the risk-based participation of this new entity in TOU-PBDRP as responsive demand and resource well has not been investigated. [136] uses a statistical model to consider the uncertain behaviour of EV owners for minimization of charging the cost of EV owners and revenue maximization of EVA for only regulation market participation with real-time prices (RTP) and load profile. The linear problem considered perspectives of SO and EV owners. Grid stability is ensured through adaptive charging rate control. [159] investigated charge scheduling problem to reduce cost from SO perspective by controlling EV charging loads in response to TOU price in a regulated market for flatting the load curve with appropriate partitioning of the peak and valley periods. However, market participation for profit maximization of EVA and storage capability through discharge scheduling is not incorporated into the model. [154-155] focus on SO perspective without addressing EVA and EV owners' concerns. [51] The day-ahead scheduling problem is investigated from the perspective of EVA's profit maximization through multiple market participation viz. energy, regulation, and reserve markets based on real-time load data and RTP. However, assessment of the management of EVA through TOU-PBDRP has not been accounted for. Besides, transformer overloading from SO perspective and charging cost reduction from EV owners' aspect has not been integrated into the model for large scale EV penetration. [70, 152, 155] incorporated TOU-PBDR and charge scheduling to reduce charging cost and flattening the load curve. Multi-market operations of EVA providing energy and regulation services through bidirectional V2G scheduling is particularly challenging due to multiple uncertainties viz. mobility behaviour and market price volatilities. However, to the best knowledge, the risk behaviour of EVA participating in multiple markets with profit variability and discharge scheduling incorporating hierarchical control and TOU-PBDR considering market price and mobility behaviour uncertainties has not been addressed comprehensively and hence requires wider investigation for improved market operations. Finally, risk-averse integrated DR and dynamic V2G scheduling of EVA in energy and regulation markets to support grid stability is modelled. The final working chapter puts forward a risk-averse decision-making algorithm to maximize the expected profits of EVA offering regulation services to SO for supporting grid stability and avoiding inevitable contingencies. For high penetration of EVs with coincident charging, becomes very problematic due to their sudden large load leading to system failure. It requires resolving system overloads and voltage deterioration by incorporating additional congestion management constraint supporting grid stability. Integrated PBDR and dynamic charging-discharging based V2G scheduling can avoid system failure by incorporating the flexible behaviour of EV owners. Hence, increased EV owners' participation in V2G service provision in response to attractive charging tariff becomes beneficial for EVA in the smart grid environment. From EV Owners' perspective, the energy tariff for charging must be lower than the upper bound of charging cost. SOC updating of an EV at time t is based on its connection hours. Proposed algorithm considering various uncertainties of energy and regulation prices using a stochastic programming framework. Pragmatic problem formulation capable of reducing battery degradation costs or increasing lifetime of EV batteries due to smooth variation of charging-discharging rates, not accelerating the battery degradation and associated financial risk. Economies of V2G operations, its resiliency and scalability to EVs' random mobility behaviour through sensitivity analysis of EVA's performance parameters concerning round trip efficiency, battery replacement cost, number of EVs, and upper limit of charging/discharging rates. Simulation results validate the efficacy and viability of the proposed business model regarding peak load management and monetary benefits for all stakeholders viz. SO, EV owners, and EVA. As a part of future work, synergistic renewables and V2G integration could be explored for sustainable development.

2.9 Epilogue

The research work presented in this thesis offers an innovative mechanism to deal with the technical aspects of electricity grids operation along with market operation through risk-averse G2V scheduling framework of price-taker EVA that aims to maximize its expected profits. It focuses on integrating DR, as well as on methods to generate revenue from this demand flexibility, by reducing the costs of procuring the charging demand, and by providing ancillary services. PBDR is integrated from EV owners' perspective of charging cost minimization and SO's perspective of valley filling. This thesis provides innovative, viable business models for enabling the integration of TOU-PBDR into the risk-averse G2V scheduling model.

Fig. 2.18 illustrates the thesis organization in nutshell with a detailed chapter outline. The theoretical background provides an insight into the role of EVA in the power system and the electricity market as a service provider. Then the base case i.e. the uncontrolled/uncoordinated V2G scheduling with adverse impacts is described. Basic Static G2V scheduling Model shows the given model and how it is designed along with further modifications made to the existing heuristic model. The case study describes the different operation scenarios and cases for which the simulations are run and then the results obtained are discussed. Validation of the proposed model describes the case studies considered and sensitivity analysis done on updating EVs penetration and upper limit of the charging rate to verify the performance parameters from the simulations. Techno-Economic evaluation evaluates the results from a technical and economic perspective and discusses the feasibility of the target operating scenario from the results obtained. Summery wraps up the chapter with the conclusions drawn from the technical and the economic analysis performed. Future Work mentions the possibilities for further studies/work.

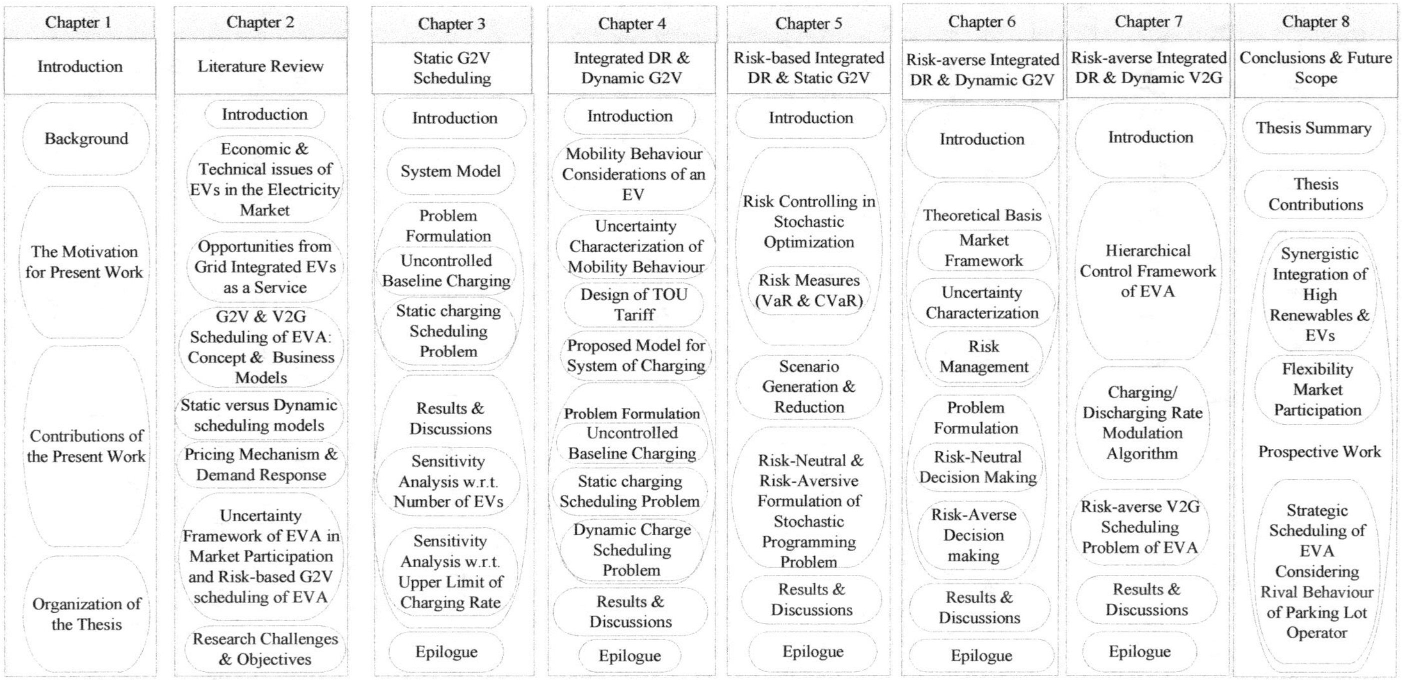

Fig. 2.18: Thesis organization depicting chapter outline

Chapter 3

STATIC G2V SCHEDULING OF ELECTRIC VEHICLE AGGREGATOR CONSIDERING GRID STABILITY AND EV OWNERS' PERSPECTIVE

In the previous chapter economic and technical issues associated with the V2G scheduling of EVA have been thoroughly discussed. This chapter is devoted to developing a static G2V scheduling algorithm of EVA considering grid stability through regulation services and EV owners' perspective of charging cost reduction. This chapter considers the static G2V charge scheduling scenario where the charging profile information from EV owners is known to EVA in advance. The proposed approach is compared with conventional unregulated G2V charge scheduling. This chapter is organized as follows. In Section 3.1, a brief overview of the G2V scheduling methods along with related work done in this field is included to maintain the continuity of the work. A detailed overview of the system architecture required to evaluate the performance parameters of EVA is presented in Section 3.2. Description of various mathematical terms involved in the problem formulation is explained in Section 3.3 while Section 3.4 illustrates the simulation results of the optimization problem and their analysis. Section 3.5 concludes the chapter.

3.1 Introduction

In a smart grid, the energy management system efficiently communicates information in an EV charging network between the power grid, EVA and EV owners. This information can be utilized by EVA to develop efficient operation strategies for intelligent load aggregation; customer costs reduction and demand satisfaction and system overloading prevention. As mentioned earlier in Chapter 2 & Section 2.5, based on available information before the G2V charge scheduling set-up is divided as static/offline charging and dynamic/online charging. In the proposed work for simplification, a static charging scenario is considered in which EVA has prior complete charging profile information from EV owners (times of arrival and departure, initial

SOC on arrival and desired SOC on departure) ahead of the present timeslot. The charging profile information may be represented as $\left(EV\ ID,\ ta_i,\ td_i,\ SOC_i^{INI},\ SOC_i^{DES}\right)$. EVA computes a charging schedule according to the information of price and next drive through static G2V charge scheduling problem (SCSP) algorithm, and individual EV owner obeys it for charging operation after its arrival.

EVA minimizes charging cost considering EV owners' perspective. However, the lowest electricity cost for charging does not guarantee optimal system performance. It is considering SO's perspective, its goals for maximizing ancillary service benefits for the grid. EVA being a commercial entity aims at revenue maximization in its self-scheduling. In other words, it restricts simultaneous minimization of the charging cost and maximization of regulation. All these problems focus on the individual perspective, disregarding other entity's concerns and does not guarantee performance. An optimally coordinated static G2V charge scheduling of EVA could resolve amalgamated concerns of SO and EV owners attaining co-optimization of techno-economic objectives, i.e. to support grid stability from regulation services and minimize charging cost as well.

For grid stability, even load distribution and frequency stability are prerequisites for SO. EVA can support in this regard to produce relatively even load distribution by applying specific POP in each hour according to baseload i.e. charge EVs with lower POP when baseload is high, and vice versa. Thus, EVA controls the speed of charging to avoid stress on the system from an increase in peak load through deviations from scheduled POP. Consequently, EVA can support balancing demand and supply by reducing the flow of power from the grid in demand upsurge condition. Proposed work represents static G2V scheduling of EVA for simultaneous optimization of EVA's revenue from supporting grid stability and EV owners' charging cost. A comprehensive study for EV charging from a customer's perspective is conducted, and an efficient strategy is developed to obtain a trade-off between the EVA's revenue and charging costs. This strategy also allows EV owners to specify their charging requirements. Besides, the static charging scenario is simple to implement as a regular and accurate method, reducing the computational burden as compared to the heuristic charging scenario in related works.

3.2 System Model

The proposed G2V charge scheduling architecture has been illustrated schematically in Fig. 3.1. This system model provides clarity on typical constituents of G2V scheduling architecture illustrating grid integration of EVs by an EVA, a substation, and a group of associated transformers. A leaf node transformer is the one that is connected to the lateral of the node to which an EV is attached. In this architecture, the aggregated EVs are connected to the leaf node transformers ensuring that the delivery capacity of each of the leaf node transformers is not exceeded by a controlled EV charging rate to prevent overloading of transmission lines and other equipment. An EVA may charge the EVs at any suitable rate of charging based on battery capacity and the load conditions, at off-peak load with higher charging rates and vice versa. Simultaneously accomplishing these responsibilities, EVA also endeavours to revenue optimization.

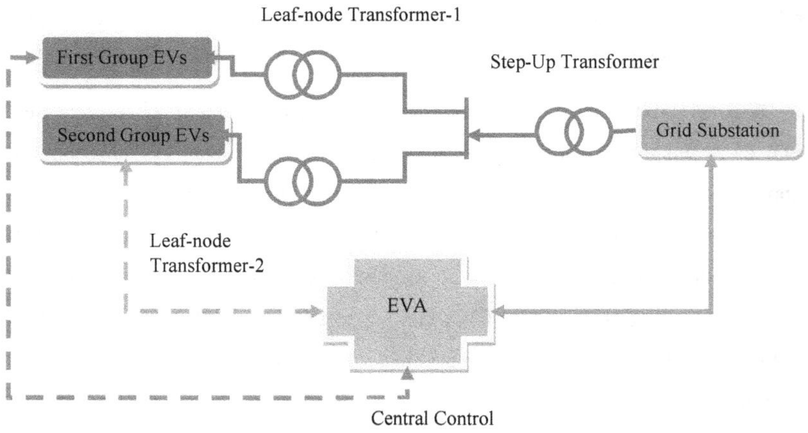

Fig. 3.1: EV Charging Network Architecture

Usually, EVA purchases power at a wholesale price from the grid, which varies from hour to hour within a day. In a dynamic pricing scenario, for peak duration times of the day, this purchase cost becomes high, and for off-peak hours, the same price becomes comparatively low. The EVA adds some price margin to this wholesale price and then sells it to the customers, to reduce its market price risk. This added price margin is known as the mark-up price. The EVA gathers information from both the SO and the EV owners. In the model considered in this study, it is assumed that an EV customer is free to arrive at any instant and depart at any moment according to their requirements.

Individual charging activity can be constituted by five sets of data, namely i index for EV number, ta_i the arrival time of ith EV, td_i departure time of ith EV, SOC_i^{INI} initial SOC of ith EV and SOC_i^{DES} desired SOC of ith EV. Although the information for some of the charging tasks may not be available in advance, such an assumption seems quite acceptable to begin with as the present EV charging schemes also do not involve that much complexity and most of them operate on advance information given a regular vehicle movement pattern. At night time, the load is low, and that is why many vehicles prefer charging in this period only. It is assumed that the considered electricity market is hourly. Henceforth, the G2V charge scheduling horizon is 24 hours spread over the whole day, from noon of the current day to noon the next day. Each interval is of one-hour duration.

3.2.1 Regulation Capacities Provided by an Individual EV

The SO has to ensure system stability and reliability preferably through flat or evened-out load distribution and a stable frequency. The EVA can deviate the charging rates of the connected EVs $POP_{i,t}$ to produce a flat or evened-out load distribution by charging the vehicles with lower charging rates during the peak periods and with higher charging rates during off-peak periods. During an emergency, if the SO calls EVA for additional power requirements, then EVA responds with temporally increased or decreased charging rates. This action is referred to as regulation service in the electricity market. The amount of charging rate that an EVA can increase or decrease in response to the grid requirement is referred to as regulation capacity.

Fig. 3.2 illustrates how EV capacity can be used for providing regulation up/down. In other words, it demonstrates the quantity of regulation capacity that a single EV may provide to the SO. The upper charging rate limit is subject to the EV battery as it is the largest rate at which the battery can draw power. The maximum charging rate limit $POP_{i,t}^{\max}$ is the power required to charge the ith EV to its desired SOC in a one-time slot t which is considered by the subsequent equation:

$$POP_{i,t}^{\max} = \frac{\left(SOC_i^{DES} - SOC_{i,t}\right) \cdot BC_i}{Ef_i} \tag{3.1}$$

Here, $SOC_{i,t}$ is the SOC level, BC_i is the battery capacity (kWh) and Ef_i is the charging efficiency of ith EV battery. The charging rate for the respective EV is not allowed to surpass the maximum charging rate limit in any time slot. Regulation up capacity $uc_{i,t}$ means when the system is having deficit generation and need to balance this amount by the reducing rate of charging to reduce the consumption from EVs load. The $uc_{i,t}$ is computed as the difference between its scheduled charging rate limit and its lower charging rate limit by which the EV can decrease the charging rate to help boost the system frequency. Regulation-down capacity $dc_{i,t}$ means when the system is having surplus generation and need to balance this amount by the growing rate of charging for i^{th} EV at a specific time t for an increase of charging energy consumption by EVs. The difference between the upper charging rate limit or maximum charging rate limit (whichever is smaller) and the scheduled charging rate is the $dc_{i,t}$. Furthermore, once charging a cluster of EVs, the overall regulation down capacity should be additionally limited by the transformer delivery capacity TDC. EVA governs whether the aggregated charging rates $POP_{i,t}$ (kW) and regulation down capacities $dc_{i,t}$ (kW) of entire plugged-in N EVs is within TDC (kW) i.e. guaranteeing $\sum POP_{i,t} + dc_{i,t} \leq TDC$. Regulation capacity $rc_{i,t}$ is thus the total of regulation down capacities $dc_{i,t}$ and regulation up $uc_{i,t}$. In this manner, EVA may support retaining the system frequency-stable offering regulation services to SO. Therefore, SO willingly approves to recompense for the regulation services offered by EVA.

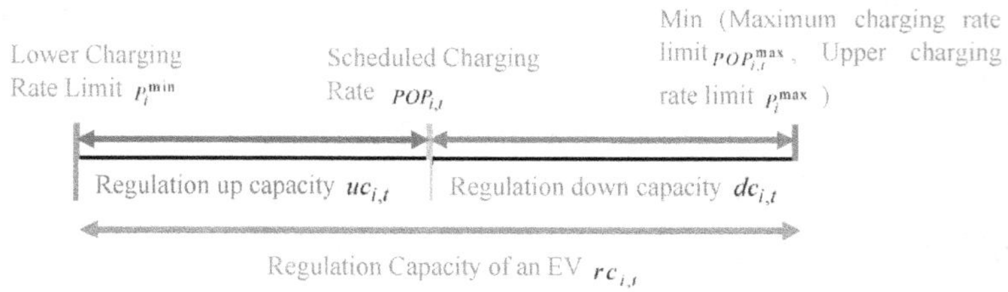

Fig. 3.2: Regulation Up/Down Capacities Offered by a Specific EV

3.3 Problem Formulation

The proposed static G2V charge scheduling problem considers two conflicting objectives: to minimize EV owners' charging cost and to maximize the EVA's revenue, constrained by different technical and economic constraints. The optimal result comprises charging schedules, regulation capacity and SOC of connected EVs, having a given charging task over the whole scheduling horizon. While maximizing the EVA's revenue, EV owners' charging cost should not exceed its upper bound. Any charging schedule will be considered feasible when both the EV owners' demand (specified by charging tasks) and the system load are satisfied.

3.3.1 Uncontrolled Baseline Charge Scheduling Problem

In the uncontrolled type of charging, EVA charges an EV in a single time slot as soon as it is plugged into the grid. Therefore, such charging is non-price responsive. In the uncontrolled charging or baseline setup, EVs are charged by the maximum rate of charging $POP_{i,t}^{\max}$ (kW).

Unregulated charging does not incorporate any economic decision-making rationale. Therefore, the total charging energy cost of unregulated charging may represent cost upper bound. This cost bound depends on the pricing scheme. An EV may not be available for a particular slot, and this would affect the computation of cost, charging energy and regulation capacities. $Av_{i,t}$ being the binary variable gives the plug-in availability of EV i in the time interval t as per Eq. (3.2). Notably, this $Av_{i,t}$ depicting EV's availability or unavailability can be either 1 when the time interval is greater than arrival time and lower than departure time; or 0 otherwise. The upper bound of charging cost (CB) is the higher value charging cost corresponding to the availability of EVs as Eq. (3.3).

$$Av_{i,t} = \begin{cases} 1 & ta_i \le t \le td_i \\ 0 & otherwise \end{cases} \tag{3.2}$$

$$CB = \sum_{t=1}^{T} \sum_{i=1}^{N} POP_{i,t}^{\max} \cdot \left(M_t + EP_t^{TOU}\right) \cdot Av_{i,t} \tag{3.3}$$

3.3.2 Proposed Static G2V Charge Scheduling Problem (SCSP)

The proposed static G2V charge scheduling problem (SCSP) considers two conflicting objectives: to minimize EV owners' charging cost and to maximize the EVA's revenue, constrained by a) system constraints and b) EV constraints, paying more attention to EV owner's satisfaction and benefit to SO. Feasible charging task satisfies both economic constraints of EV and technical constraints of the system. The optimal result comprises charging schedules, regulation capacity and SOC of connected EVs, having a given charging task over the whole scheduling horizon. While maximizing the EVA's revenue, EV owners' charging cost should not exceed its upper bound. The proposed problem can be defined via two statements:

Problem 1: For the set of given charging tasks, the EVA's revenue $\sum_{i=1}^{N}\sum_{t=1}^{T} REV_{i,t}^{SCSP}$ should be maximum among all the feasible charging tasks in SCSP.

$$Rev^{SCSP} = \sum_{t=1}^{T}\sum_{i=1}^{N} RP_t \cdot rc_{i,t} + \sum_{t=1}^{T}\sum_{i=1}^{N}\left(M_t + EP_t^{RTP}\right)\cdot POP_{i,t} \cdot Av_{i,t} \tag{3.4}$$

Here, for i^{th} EV at time slot t, RP_t is the regulation price (\$/kWh), M_t is mark-up price (\$/kWh) and $Av_{i,t}$ is indicating whether EV is available/connected or not. The revenue of EVA comprises of two terms as shown in Eq. (3.4). First is the regulation service offered to the SO. EVA provides both regulation services (up and down) to SO. For the considered symmetrical market New York Independent System Operator (NYISO), EVA sells both the services at the same price. Therefore, revenue from regulation is a product of regulation power price at the time t and total regulation capacity offered by EV i at a time t. Further EVA procures energy from the wholesale market at a real-time price EP_t^{RTP} and sells it at a mark-up M_t over EP_t^{RTP}. Thus, the second term implies revenue from selling energy to satisfy charging requirements of available EVs, as a product of mark-up price over EP_t^{RTP}, charging rate of EV i at a time t.

Problem 2: For the set of given charging tasks, the EV owners' charging cost $\sum_{i=1}^{N}\sum_{t=1}^{T} CC_{i,t}^{SCSP}$ should be minimum among all the feasible charging tasks in the SCSP.

$$CC^{SCSP} = \sum_{t=1}^{T} EP_t^{RTP}\left(\sum_{i=1}^{N} POP_{i,t} \cdot Av_{i,t}\right) \tag{3.5}$$

As mentioned earlier, for the specified charging tasks set, the revenue of EVA must be highest amongst all viable tasks of charging in contradiction to the purchase cost CC^{SCSP} which as per Eq. (3.5) should be minimum. The present work formulates SCSP as a profit maximization problem of EVA as a difference in its revenue and costs to quantify the dynamics of the problem. Therefore, the scheduling problem can be defined via the statement (3.6):

$$\max_{SOC_{i,t},\mathrm{POP}_{i,t},rc_{i,t}} \quad \Pr ofit = \mathrm{Re}v^{SCSP} - CC^{SCSP} \tag{3.6}$$

In a static charging scenario, EVA has complete knowledge of charging profile information of ith EV: ta_i, td_i, SOC_i^{INI}, and SOC_i^{DES}. Arrival time ta_i and departure time td_i are predicted from the history of the mobility pattern of the ith EV. It is assumed that they follow the Gaussian probability distribution function (pdf) as given in Eqs. (3.7) and (3.8). Here, μ_{ta_i} and μ_{td_i} are the mean, whereas σ_{ta_i} and σ_{td_i} are the standard deviation of the normally distributed ta_i and td_i, respectively.

$$pdf(ta_i) = \frac{1}{\sigma_{ta_i}\sqrt{2\pi}} \exp\left(-\frac{(ta_i - \mu_{ta_i})^2}{2(\sigma_{ta_i})^2} \right) \tag{3.7}$$

$$pdf(td_i) = \frac{1}{\sigma_{td_i}\sqrt{2\pi}} \exp\left(-\frac{(td_i - \mu_{td_i})^2}{2(\sigma_{td_i})^2} \right) \tag{3.8}$$

In this constraint (3.9), we assign a value to the SOC of each connected EV in each hour. $Av_{i,t}$ is availability status for i^{th} EV during a time slot t. This availability of EVs would have a significant effect on the calculation of regulation capacities, charging energy and cost.

$$SOC_{i,t} = \begin{cases} SOC_i^{INI}, & \text{if } t = ta_i \quad \forall i \\ SOC_i^{DES}, & \text{if } t = td_i \quad \forall i \\ SOC_{i,t-1} + Ef_i Av_{i,t-1} POP_{i,t-1} / BC_i, & \text{otherwise} \end{cases} \tag{3.9}$$

EVA is required to fulfil the charging requirement of EVs. Therefore, at the departure hour, the charging rate of an EV i is taken as deadline constraint (3.10),

$$POP_{i,t} = \left(SOC_i^{Des} - SOC_{i,t-1} \right) \cdot \frac{BC_i}{Ef_i} \cdot Av_{i,t} \quad \forall \mathrm{t} \in \mathrm{td}_i \tag{3.10}$$

$$P_i^{\min} \le POP_{i,t} \le P_i^{\max}, \qquad \forall i,t \tag{3.11}$$

$$\sum_{t=1}^{T} (M_t + EP_t) \left(\sum_{i=1}^{N} (\mathrm{POP}_{i,t}\, Av_{i,t}) \right) \le CB \tag{3.12}$$

$$rc_{i,t} = uc_{i,t} + dc_{i,t} \qquad \forall i,t \tag{3.13}$$

$$uc_{i,t} = \begin{cases} \left(POP_{i,t} - P_i^{\min} \right) \cdot \mathrm{Av}_{i,t} & \forall i,t \;\; s.t. \;\; \mathrm{Av}_{i,t} = 1 \\ 0, & \text{otherwise.} \end{cases} \tag{3.14}$$

$$dc_{i,t} = \begin{cases} \left(\min\left\{POP_{i,t}^{\max}, P_i^{\max}\right\} - POP_{i,t}\right) \cdot \mathrm{Av}_{i,t} & \forall i,t \ \ s.t. \ \mathrm{Av}_{i,t} = 1 \\ 0, & \text{otherwise.} \end{cases} \quad (3.15)$$

$$\sum_{i=1}^{N} (\mathrm{POP}_{i,t} + dc_{i,t}) + l_t^b \leq TDC \qquad \forall \mathrm{t} \quad (3.16)$$

In constraint (3.11), for $POP_{i,t}$ charging rate of ith EV in a time slot t, $P_i^{\min}$ is lower charging rate limit (kW) and $P_i^{\max}$ is the upper charging rate limit (kW). In constraint (3.12), CB is the upper bound of the charging cost ($) of an EV. It is the total charging cost of an unregulated scenario, where the vehicles are charged with the maximum possible charging rate as depicted in (3.3). Constraints (3.13-3.15) are based on the system model, detailed in the previous section, to calculate the regulation capacities.

As mentioned in earlier section 3.2, the overall regulation capacity $rc_{i,t}$ is equivalent to the sum of the regulation up $uc_{i,t}$ and regulation down capacities $dc_{i,t}$ represented by (3.13). Regulation up/down capacity for i^{th} EV at a specific time t, $uc_{i,t}$ / $dc_{i,t}$ are respectively computed using by Eqs. (3.14) and (3.15). EVA determines in each time slot t, whether the summation of charging rate $POP_{i,t}$ (kW) and regulation down capacity $dc_{i,t}$ (kW) for all connected EVs N is within the distribution transformer delivery capacity TDC (kW). In this way, it guarantees that the EVs are charged within the permissible range of transformer delivery capacity and avoid overloading of the transformer caused by the charging of EVs. This congestion management constraint is expressed as Eq. (3.16), ensuring that TDC is always respected. Here, l_t^b is the baseload at the time slot t. It should be noted that since the assumed market for regulation service is hourly, any EV which is not available for a particular hour will not contribute in the total regulation service provided by the EVA to the SO.

This study describes mathematical optimization applied to maximize (3.6) subjected to constraints (3.1) -(3.3) and (3.10) - (3.16). Here, for a time slot t, CC^{SCSP} is the charging cost of all EVs ($) and EP_t is the electricity cost ($/kWh). It maximizes (3.4) and minimizes (3.5) as well. The detailed mathematical formulation and the results are described in the following sections. To perform multi-objective optimization of objectives given by equations (3.4), (3.5) and for attaining a trade-off solution, an optimization framework is set up in Matrix Laboratory (MATLAB). Analyzing the objective functions and the constraints, it can be seen that the complete optimization problem is mixed-integer linear due to the incorporation of binary variable and continuous variables. It can be simplified to a linear problem if the availability is not considered as a binary variable instead predetermined based on communicated information of arrival and departure times. Hence, the Linear Programming Problem (LPP) is solved using the Simplex method in MATLAB and results are detailed in the next section.

3.4 Results and Discussions

The proposed static G2V charge scheduling scheme is evaluated for actual electricity market price and actual baseload and battery data. Precisely, electricity and regulation prices are obtained from the hourly market NYISO [134]. The average of the electricity and regulation prices over 30 days (Jan 1-30, 2015) is used for simulation. The realistic data of μ_{ta_i}, μ_{td_i}, σ_{ta_i} and σ_{td_i}, and TDC are captured from [93]. The base loads at each hour are taken from the load profile in [134] while the battery capacity efficiency, charging rate limit are considered referring [148]. Different constants used in simulations are specified in Table 3.1. The EVA is assumed to charge a mark-up price $M_t = 0.05$ \$/kWh from EV owners [149]. The SOC_i^{INI} is set to be distributed in the range of [0.3, 0.9] while SOC_i^{DES} is set at 0.9 for each EV. The charging rate $POP_{i,t}$ varies in the range of 0 to 4.4 kW. A total of ten Monte Carlo simulations are used to generate ten sets of random input data, and the average values over ten runs are presented as results in the following figures. The performance parameters of EVA: EVA's revenue and the total charging costs of the EVs are calculated for the optimal trade-off solution. To analyse the impact of the number of EVs N and the upper charging rate limit $P_i^{\max}$ on the performance parameters, simulations are done for different values of N and $P_i^{\max}$ for comparison.

Table 3.1: Simulation Parameters

Mean of t_{a_i}	**Mean of t_{d_i}**	**Standard Deviation of t_{a_i}**	**Standard Deviation of t_{d_i}**
7	19	2 hours	2 hours
T	TDC	E_{f_i}	BC_i
24	200 kW	0.9	16 kWh

3.4.1 Sensitivity Analysis of Performance Parameters of EVA with the Number of EVs

The grid integration of EVs is based on the vehicle stock in a region. If the number of EVs in a region does not reach a critical value, most of the concerned entities are not willing to put in place EVs G2V charge scheduling management systems and services. In other words, if the number of pluggable EVs is low, the setup of EV battery loading services is not justified from the cost-benefit point of view: efforts and costs are fixed regardless of the number of plugged vehicles, while benefits depend on vehicle stock. Thus, the number of EVs indicate penetration level out of conventional vehicles stock. Fig. 3.3 depicts the performance indices of EVA in terms of monetary values i.e. revenue, charging cost, and profit with varying number of EVs from 5 to 200 for an upper charging rate limit of 4.4 kW optimizing Eq. (3.6). It is observed that with the increase in the number of vehicles, the minimum total charging cost for customers is decreasing, i.e., with a large number of customers having EVs the overall charging cost per customer significantly reduces. Also, the EVA gets more revenue with a large number of EV customers.

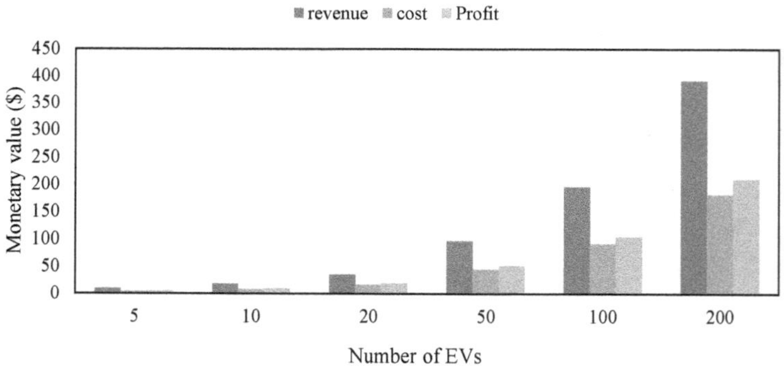

Fig. 3.3: Performance Parameters of EVA Versus the Number of EVs

Figs. 3.4 and 3.5 show the comparative analysis of monetary values with the number of EVs in G2V charge scheduling for different methods i.e. baseline (unregulated) case and proposed method after optimizing Eqs. (3.5) and (3.4) respectively. It validates the efficacy of the proposed method. It is clear from the graphs that the proposed scheme with the simplex solution approach gives better results. Also, the regulated charging scheduling gives significantly lesser minimum total charging cost and larger maximum EVA's revenue as compared to uncontrolled baseline charging of the EVs. Fig. 3.4 also summarize that with the increase in the number of vehicles, the minimum total charging cost for customers is decreasing, i.e., with a large number of customers having EVs the overall charging cost per customer significantly reduces. From Fig. 3.5 we can say that the EVA also gets more revenue with a large number of EV customers.

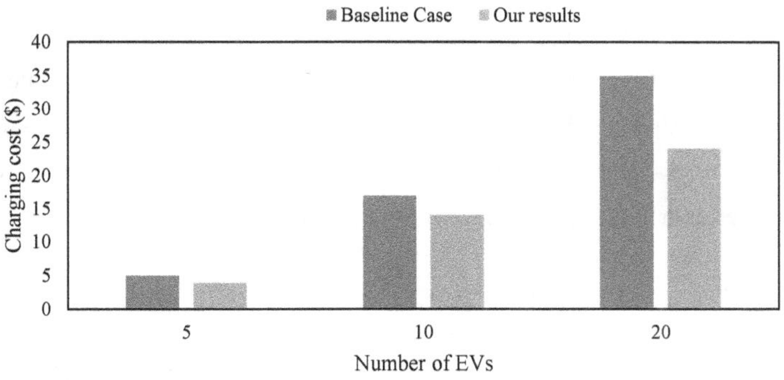

Fig. 3.4: Minimum Total Charging Cost Versus the Number of EVs

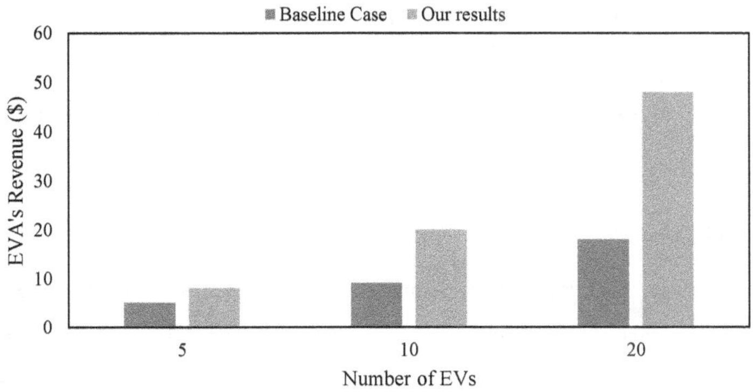

Fig. 3.5: Maximum EVA's Revenue Versus Number of EVs

3.4.2 Sensitivity Analysis of Performance Parameters of EVA with Upper Charging Rate Limit

The impact of the upper charging rate limit on performance indices of EVA in terms of monetary values i.e. maximum EVA's revenue, the minimum total charging cost, and profit is shown in Fig. 3.6. Simulation runs are conducted for $N = 200$ EVs by changing the upper charging rate limit from 3.3 kW to 8.8 kW with a step size of 1.1 kW. It can be seen from Fig. 3.6 that a rise in the upper charging rate limit does not cause significant benefit for EV owners. Though, maximum EVA's revenue increases almost linearly with the upper charging rate limit because this limit improves the regulation capacity. Hence, increasing the charging rate can only be made possible and realized effectively by improving the EVA framework. However, a considerably higher value of charging rate may increase the system peak load demand due to EVs charging, result in overheating, thus increase system losses. For a smaller group of EVs $N = 5$, the impact of the upper charging rate limit on maximum EVA's revenue and minimum total charging cost is shown in Figs. 3.7 and 3.8, indicating that a rise in the upper charging rate limit does not cause significant benefit for EV owners. Nevertheless, maximum EVA's revenue increases almost linearly with the upper charging rate limit.

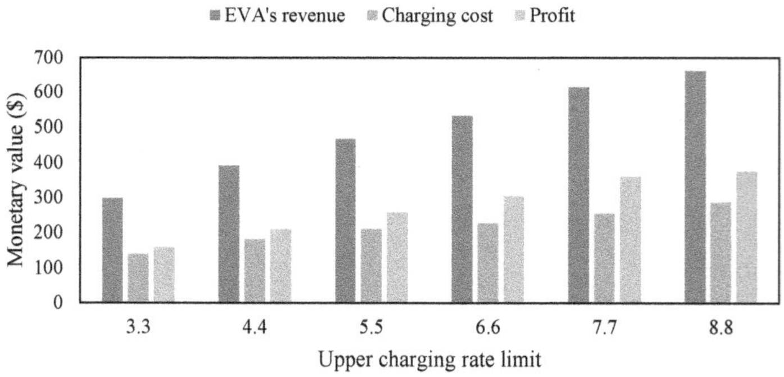

Fig. 3.6: Performance Parameters of EVA Versus Upper Charging Rate Limit

Furthermore, EV charging can create new load peaks exceeding the service transformer's rated delivery capacity, thereby increasing the daily expansion and contraction of the transformer, resulting in a negative effect on the transformer's life. It causes accelerating equipment ageing and an economic burden on SO. Thus, transformer overloading is an important issue to slow down the growth of the higher value of the charging rate. Through smart coordination with SO, EVA receives the value of transformer delivery capacity and respects loading limits accordingly for G2V scheduling.

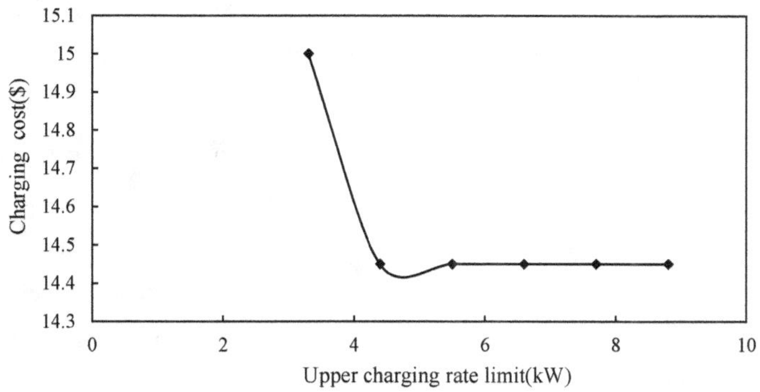

Fig. 3.7: Minimum Total Charging Cost Versus Upper Charging Rate Limit

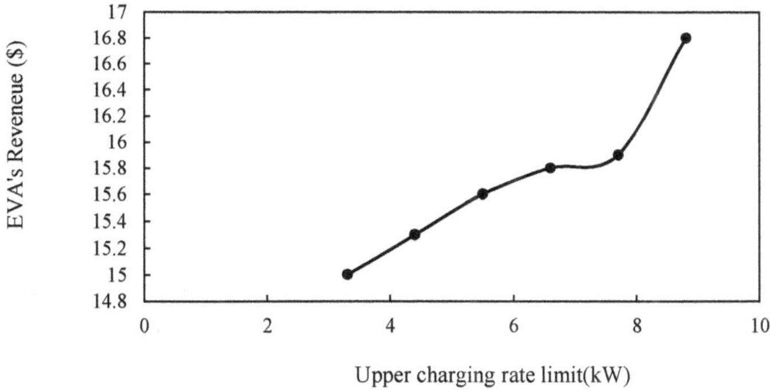

Fig. 3.8: Maximum EVA's Revenue Versus Upper Charging Rate Limit

3.5 Epilogue

Considering the potential of coordinated regulated EV charging, this chapter proposes static G2V scheduling of EVA to decrease the potential impacts of new load peaks created by large EVs integration, accommodate EV charging while keeping the peak demand unchanged, and maximizing the grid usage. A mathematical optimization problem for the development of a regulated EV charging scheme has been formulated considering the EV owners' cost along with EVA's revenue. A static charging scheduling scenario is considered. The LPP has been solved using the simplex method in MATLAB. Simulation results based on real electricity price and load data are presented to show the revenue gains and cost savings brought by optimal static G2V charge scheduling. Comparative analysis of the EVA revenue and the charging costs is carried out for uncontrolled and controlled charging scenarios. Additionally, their sensitivity towards the different number of vehicles as well as upper charging rate limit is analysed. Results show interesting trends regarding the economics of EV charging, grid stability and charging rate control.

EVA's revenue and total charging cost are quantifiable measures determining the behaviour and performance of optimizing the algorithm. A kind of business metrics used to track and assess the status of the optimality process. Optimal solutions given by static charging scheduling schemes can be used to show the potential benefit that can be brought by regulated charging and serve as a benchmark for performance evaluation. Significant revenue gains and cost savings are brought by optimal charging scheduling as compared to the unregulated baseline case. Since the charging cost is lesser than the above-mentioned case. Also, the profit of EVA is more as compared to the baseline case.

The total charging cost can be reduced by 18.5% and EVA's revenue can be improved by 139% on average in comparison to the unregulated baseline approach. Hence, the proposed approach is better as the number of EVs increases, EVA's revenue increases at the same time charging cost becomes cheaper satisfying EV owner requirement as well. Hence, the proposed work is beneficial to lower network congestion and increased grid support regulation services for SO. Results are better reflecting customer perspective by reducing charging cost burden as well.

This chapter gives a brief idea of incorporating a customer perspective to an existing charge scheduling approach. Pragmatically the charge scheduling plan of EVs is prior not stated to EVA, and EV owners are flexible for arrival/departure whenever they desire. It requires a more realistic local/online/dynamic G2V charge scheduling problem (DCSP) set-up in which the EVA does not have EVs information before their actual arrival. As a part of future work, the problem can be converted into a more realistic case for the dynamic charge scheduling problem. Besides, EV owners charging cost can be reduced further using demand-side management programs. Thus, incentivizing EV owners is further propagated, to reflect their contribution to regulation services. EV mobility uncertainty under dynamic scenario incorporated in the problem formulation increases the computational burden. Therefore, appropriate stochastic optimization algorithms are required. The next chapter will deal with all these issues.

Chapter 4

INTEGRATED DEMAND RESPONSE AND DYNAMIC G2V SCHEDULING OF EVA CONSIDERING GRID STABILITY

In the previous chapter, static G2V scheduling of EVA is proposed where the EV owner perspective is emphasized. To motivate EV owners further to participate in coordinated G2V scheduling, this chapter integrates TOU-PBDR into the G2V scheduling model. Besides, a pragmatic scenario for the dynamics of mobility behaviour is considered in this chapter. Benefits to EV owners are enhanced due to the offering of a special designed TOU tariff to them. The consequences of these dynamics are assessed from the perspectives of all involved entities.

The chapter is organized as follows. The background and motivation behind the work are illustrated in Section 4.1; mobility behaviour considerations of an EV and uncertainty characterization of mobility behaviour for an EV are described respectively in Sections 4.2 and 4.3. Section 4.4 details the design of the TOU price to be offered to EV owners for charging EVs. The proposed model for the system of charging and the EVs charge scheduling optimization algorithms that have been used are articulated in Section 4.5. Problem formulation is discussed in Section 4.6. A comparison of the results of the proposed model Dynamic Charging Scheduling Problem (DCSP) with the existing methods Static Charging Scheduling Problem (SCSP) and Uncontrolled Baseline Charging is discussed in the successive sub-sections. Results are analysed in Section 4.7. The outcomes of mathematical optimization and their concluding remarks are realized in Section 4.8 along with the further scope of the proposed work.

4.1 Introduction

EVA can adopt coordinated and distributed charging rates for EVs over the connection hours providing stability and regulation to SO. However, against flat rate, PBDR integrated scheduling motivates EVs to charge in off-peak hours and utilize their flexibility for load levelling and peak reduction. PBDR reduces EV charging cost as well. RTP-DR, mostly adopted in literature, has a low acceptance rate by EVs for being too dynamic to respond and being infeasible for few charge cycles in a day. TOU, being less dynamic, is a natural price signal for EVs under PBDR but may reduce EVA's profit. Considering this, EV specific TOU is designed from RTP using the AHC method. In the previous chapter, buying and selling operations were on RTP wholesale electricity prices. In this chapter, buying/procurement operation is at RTP while selling to EV owners is at TOU rates specially designed by EVA.

In this context, this chapter presents optimal coordinated and dynamic G2V charge scheduling of EVA for the energy and regulation market with integrated TOU-PBDR for EV owners. EVA profit maximization is modelled as a two-stage problem, considering EV mobility dynamics, grid constraints, and EV owners' perspective of charging cost minimization as well. In the first stage, EVA designs TOU prices for the EV owners intending to successfully convert forecasted RTP to TOU without compromising the EVA revenue from customers. The conventional three-tiered TOU structure is updated to a five-tiered structure for more flexible options to EV owners and designed optimally using the AHC method. The method is adopted from [78] in this chapter. In the second stage, optimal scheduling of EVA for the energy and regulation market is conducted based on the TOU rates of stage 1 for EV owners. This stage decides the optimal POP (charging rates of the EVs) considering revenue of EVA from the regulation market and energy sell to EVs for charging, cost of EVA for energy purchase at RTP, and charging cost minimization of EVs. Also, it is assumed that the probability of SO call for regulation services is one. Through optimal TOU prices design, the proposed model has the potential to select a feasible time of EV charging operation as per their price-responsive behaviour. As discussed in Chapter 3, distribution transformer delivery capacity constraint is considered for congestion management. Monte Carlo simulation (MCS)–based scenarios generation approach is used for modelling uncertainties associated with EVs mobility behaviour, i.e. initial SOC, times of arrival, and departure [87]. A large number of these scenarios is reduced using the Kantorovich distance (KD) based backward reduction algorithm [92]. How EV owners react to offered TOU price (valley filling) and what will be the overall charging load (aggregated charging rates of all EVs at time t) to SO, the up and down regulation provided in each hour for supporting grid stability are found out as outcomes of the integrated scheduling problem. The proposed integrated TOU-PBDR and dynamic G2V charge scheduling model results in satisfying energy requirements of EVs at reduced charging cost, generate revenue to EVA and flattening the load profile of SO while reducing transformer ageing by respecting its delivery capacity constraint. Sensitivity analysis for EVA profit and EV charging cost has been conducted based on the variation in the upper charging rate limit of the EVs and the number of EVs.

4.2 Mobility Behaviour Considerations of an EV

In this work, EV mobility behaviour is characterized based on its time of arrival and departure; and initial SOC as depicted in Fig. 4.1. Other factors may be charging duration, charging power, travelled distance and recharge probability as taken up by different works of literature. However, EV charging duration is dependent on the arrival and departure time of EVs; hence, it is implicit in both [50]. Similarly, EV charging power depends on the difference of desired SOC and initial SOC, and on charging duration [128-129]. The travelled distance by EV affects the SOC of the battery, i.e. if the distance travelled by EV is more, then the initial SOC at the time of arrival will be low and vice versa. Therefore, the initial SOC reflects the distance travelled [128]. Considering recharge probability and recharging frequency per day, they depend on the penetration of EVs and the number and geographical coordinates of adjoining installed charging stations [130].

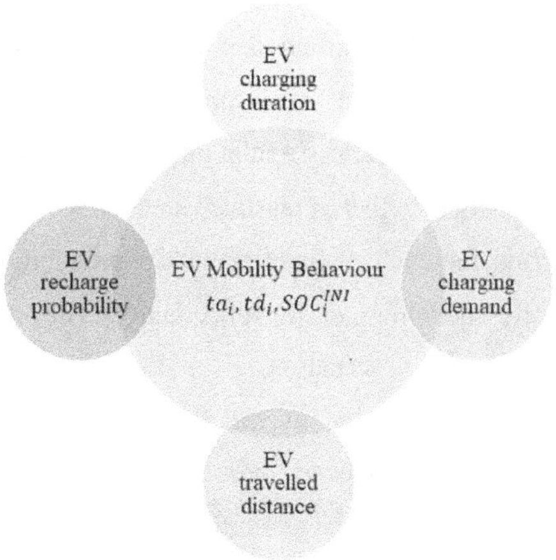

Fig. 4.1: Illustrating Stochastic Mobility Behaviour of EVs

Type and purpose of the trip in routine driving requirement value differ among weekdays and weekends and different involved seasons. In general, EV battery consumption/battery depletion would be higher in the summer season due to high ambient temperature and additional air conditioning that consumes more energy than heating. Also, flexibility and willingness to postpone charging need is dependent on plug-in time, mileage, charging power rate, incentive/ charging price structure, etc. [131]. In this work, the authors have considered weekdays of the summer season and EVs leave for morning trip with SOC 0.9 and return in the evening for recharge with an initial SOC. A standard deviation of 2 hours considers Early/late departure/ arrival due to some unexpected hurdles in its running behaviour. EV comes early for a recharge if its battery SOC reaches close to the initial SOC earlier. So, we have considered that EV has to recharge in the evening before setting out for the next trip. Thus, the recharge probability for each EV is assumed to be 1.

The work proposed replicates the online scheduling of EVs in electricity networks carried out by EVA [67-68]. EVA knows only the past and current casual information (charging demand and departure deadlines of the EVs that arrive at or before the current time, the past and current baseload from SO). Thus, EVA schedules the EVs in each hour according to the availability of EVs.

4.3 Uncertainty Characterization of Mobility Behaviour for an EV

To model realistic conditions, EV mobility parameters (initial SOC, times of arrival and departure) are considered uncertain and are assumed to follow the Normal Probability Distribution Function (pdf) with forecasted value as mean and a standard deviation for volatility. Monte Carlo Simulation is used to model these uncertain parameters in terms of a large number of feasible values as scenarios based on their pdf. One thousand scenarios for each parameter, with a probability of occurrence, are generated for complete modelling of the uncertainty. The expected value of the proposed problem over these scenarios gives the optimal solution. However, consideration of this complete set makes the problem intractable with high computational times. A powerful scenario reduction technique is essential to recover the tractability of the problem while best retaining the essential features of the stochastic information in the initially generated scenarios. It seeks to find out a scenarios' subset, and a probability measure based on this subset is required, which is closest to the initial probability distribution in terms of probability metrics. In stochastic optimization problems, the probability metrics-based scenario reduction method is more fundamental and appealing. One of the most common probability distances is the KD method, which is defined between two probable distributions. It is based on two algorithmic methods: Backward Reduction and Forward Selection. For a large number of generated and preserved scenarios, the forward selection method requires higher CPU times while backward reduction algorithms are faster and efficient. KD based method finds a trade-off between computational speed and accuracy [90, 92, 105, 110, 132]. This paper utilizes KD based backward method and reduces each of the 1000 scenarios of uncertain parameters to tractable ten scenarios. The flowchart of the scenario generation and reduction for an uncertain parameter is shown in Fig. 4.4 and the mathematical steps of scenario reduction is given below. In this method, for an initial scenario set Υ with a probability distribution Z, a new reduced scenario subset $\Upsilon_S \in \Upsilon$ of prescribed cardinality N_{Υ_S} or accuracy is to be determined and assign new probabilities to the as Z' nearest to Z in terms of probability distance. The probability distance KD can be defined as (4.1)

$$KD\left(Z,Z'\right)=\sum_{\omega\in\Upsilon/\Upsilon_s}\pi_\omega \min_{\omega'\in\Upsilon_s}\left(\upsilon\left(\omega,\omega'\right)\right) \tag{4.1}$$

Here, π_ω is the probability of scenario subset ω and $\upsilon\left(\omega,\omega'\right)$ is a function of cost defined as distance vector between scenario subsets ω and ω' evaluated using the following expression (4.2):

$$\upsilon\left(\omega,\omega'\right)=\left\|P_\omega-P_{\omega'}\right\|,\qquad \forall\omega,\omega'\in\Upsilon_S \tag{4.2}$$

P_ω and $P_{\omega'}$ are the values of scenarios. Equation (4.1) can be used to derive several heuristics for generating reduced scenario sets, which are close enough to the original set. In backward reduction, optimal deletion of a single scenario is recursively repeated until a prescribed number of scenarios are deleted. The detailed algorithm is given below.

Step1: Scenarios are generated by the MCS method as described. Assign probabilities to scenarios such that the sum of probability of all scenarios at any time step is unity. The probability of each scenario belonging to a set Υ with cardinality N_Υ is $1/N_\Upsilon$.

Step 2: Compute KD Matrix: Compute the cost matrix for each pair of scenarios using (4.2) and then determine the KD matrix by multiplication of scenario probabilities and cost matrix using (4.1).

Step 3: Scenario Selection: Find the scenario with minimum KD and mark in the KD matrix.

Step 4: Scenario Elimination: Selected minimum KD scenario is eliminated based on relative closeness to other scenario and low probability of occurrence. The deleted scenario has zero probability of occurrence.

Step 5: Update Probability Matrix: Probability of deleted scenario is added on the probability of the nearest scenario identified in previous steps. Update the initial probability matrix with a new matrix.

As stated, the uncertain variables ta_i, td_i, SOC_i^{INI} are normally distributed and independent events. Scenario generation and reduction of these for all N EVs is done individually and the scenarios of each have unique probabilities. A combined N_s scenario of the variables is created where any scenario s can be shown as Eq. (4.3)

$$\text{Scenario } s = \left\{ta_{i,s}, \cdots, ta_{N,s}, td_{i,s}, \cdots, td_{N,s}, \mathrm{SOC}_{i,s}^{INI}, \cdots, \mathrm{SOC}_{\mathrm{N},s}^{INI}\right\} \tag{4.3}$$

Each such scenario has a joint probability π_s based on the probability of constituent terms. The correlation coefficient of these terms shows a weak correlation, providing a lack of inter- dependability among them and thus, the joint probability of occurrence of any scenario s as in Equation (4.4) is obtained as a product of probabilities of constituent terms.

$$\pi_s = \prod_{i \in N} \left(\pi_{s-ta_i} \times \pi_{s-td_i} \times \pi_{s-SOC_i^{INI}} \right) \tag{4.4}$$

Here π_{s-ta_i}, π_{s-td_i} and $\pi_{s-SOC_i^{INI}}$ are the probability of scenarios of, ta_i, td_i and SOC_i^{INI} respectively.

4.4 Design of TOU Tariff

There exist many methods for cluster analysis as depicted in Fig. 4.2. AHC Method is used to determine TOU for PBDR from EVs. AHC Method emerges as the most effective algorithm generating wide clusters by comprehending the price pattern [83]. For the small-sized dataset, it is preferred due to inherent accuracy. Other approaches tend to spread price patterns among any groups generated during clustering development and, hence, are weakly competent [83, 133]. K-means outcomes could diversify depending on the choice of irregular seeds. Some random seeds may result in poor convergence rate or concurrence to locally optimal clustering. Hierarchical clustering is identified by the development of a chain or tree diagram. AHC Method initiates with each item in a discrete cluster. Thereafter, clusters are designed by an alliance of items into larger and larger groups. Cluster exploration is executed on the same data via diverse distance measures. The distances at which groups are united (from accumulation schedule) can be exhausted by the number of clusters, based on which the average within-group distance is computed. Lesser average within-group distance specifies a higher resemblance inside the group. The outcomes through these distance measures are compared based on the stability of results. The similarity grade is the Euclidean distance. For an individual group, the sum of squares is determined. The two groups with the least rise in the total sum of squares inside cluster distances are united [78, 131].

In this work, five clusters are determined for a five-tier TOU from the RTP. The aim is to convert RTP to TOU without compromising the revenue EVA. According to different types of the season 'm': summer/winter and different day types 'n': weekday/weekend, there can be four possible different RTPs and associated TOUs: summer weekday TOU, summer weekend TOU, winter weekday TOU and winter weekend TOU. Elaborated DR considering each category may be realized for the long-term planning model of EVA. The proposed short-term model considers RTP based on forecasted data for the summer season and weekdays (September 2018), from the New York Independent System Operator (NYISO) [134]. From this data, the mean and standard deviation for each hour is calculated. Then using these values, 1000 scenarios are generated based on normal pdf. Following the algorithm of Fig. 4.4, the scenarios are reduced to 10. The average hourly value over these ten scenarios is considered as expected RTP_{mn}, which is set as an initial value for the TOU price design algorithm.

For particular summer season and weekday type, this initial RTP can be represented as 1×24 a vector$\left(y_{1,}y_2,\cdots,y_{24}\right)$, according to twenty-four hours. All these RTPs are considered individual clusters. Thereby, Euclidean distances $d_{k,j}$ between two prices y_k and y_j at an hour k and j can be estimated by (4.5) as

$$d_{k,j}=\left\|y_k-y_j\right\|=\sqrt{\left(y_k-y_j\right)^2} \tag{4.5}$$

Then grouping of the hours into ladder-like tiers is obtained by amalgamating the hours with the minimum distances on their prices, hence forming a group of updated clusters. Subsequently, a distance is calculated between updated clusters to form new clusters using Ward's distance method using Eq. (4.6). Here, for any clusters P and Q, y_p and y_q are prices of hours which were assembled into them, respectively. The procedure of creating updated clusters is repetitive until five clusters remain.

$$d_{P,Q} = \frac{1}{|P||Q|} \sum_{y_p \in P} \sum_{y_q \in Q} \sqrt{\left(y_p - y_q\right)^2} \tag{4.6}$$

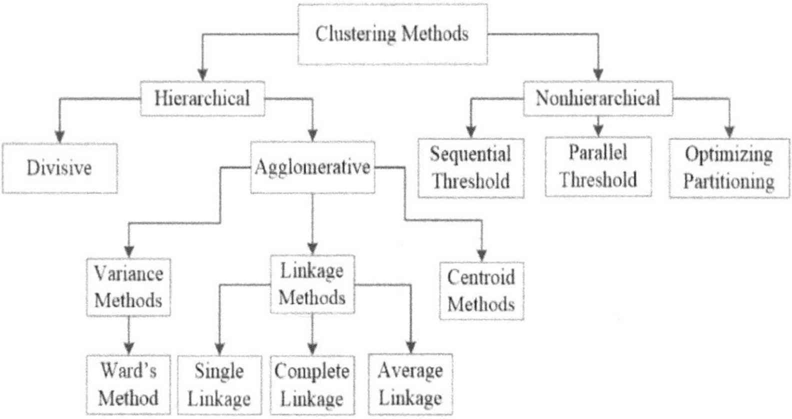

Fig. 4.2: Schematic of Cluster Analysis Methods

Considering the summer weekday under case study, each hour can be therefore assigned into one of the five clusters, i.e. off-peak, shoulder off-peak, flat, shoulder peak and peak. Fig.4.3 illustrates the dendrogram plot of considered RTP data. The numerals along the horizontal axis signify the indices of prices in the initial data set. The relations amongst prices are denoted as upside-down U-shaped lines. The height of this U indicates the distance between prices. Thus, the height of the y-axis reflects price distances [78]. After getting the clusters, the result is only the length of each cluster, but no information about the price rate of the cluster is there. Taking the average of the RTP of the hours in each cluster is most convenient. To have a total energy cost of hours within a cluster be the same, either by TOU or RTP, the tariff rate of a cluster may be determined as Eq. (4.7):

$$Height\ of\ each\ TOU\ period\left(TOU_{mn,\mathrm{K}_c}\right) = \frac{\sum_{t \in T_{K_c}} RTP_{mn,t}}{Number\ of\ hours\ in\ K_c} \tag{4.7}$$

Here, m is the season and n is day type. K_c is the cluster whose tariff is determined, T_{K_c} is the group of hours in K_c, t is the index of an hour. In the dynamic case where charging load information is not known to EVA in advance, the conventional approach to determine the height of each TOU period can be adopted as a ratio of the sum of RTPs in a tier to the tier duration. A five tiered-TOU (Fig.4.5) is devised for EV owners to provide more flexibility to them and improve the performance. Off-peak hours indicate a bottom request of energy and the smallest inflated generation sources are utilized. Flat-hours illustrate mild request for energy and the energy price. Shoulder-off peak hours characterize the energy request, and prices are inferior to a flat duration nevertheless advanced than off-peak hours. Shoulder peak hours depict the energy request and price are inferior to peak hours nevertheless advanced than flat hours. Peak-hours illustrate maximum energy request and further luxurious methods of power generation are utilized.

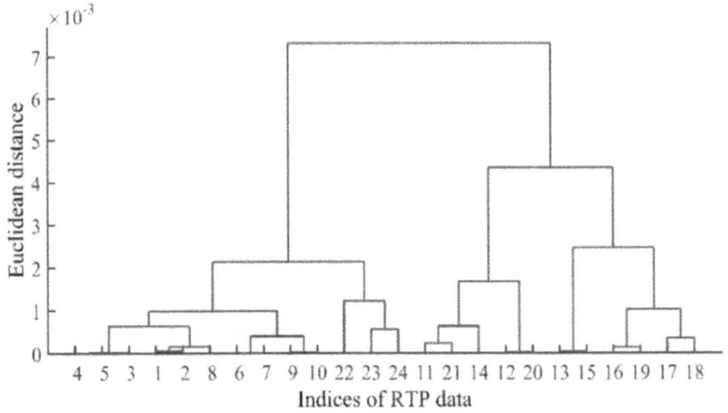

Fig. 4.3: Dendrogram Plot of RTP Data for the Weekday of the Summer Season

The focus of the problem is to execute TOU-PBDR and dynamic charge scheduling of EVs, considering a fixed market share of large EVA. EVA, aiming at its profit maximization, executes TOU-PBDR and conveys the EV owner a price-motivation to adapt its consumption to off-peak price hours of the day. Thus, there is not at all possible for EV owners to swap to a different EVA due to weak competition, information scarcity, and switching complexity, [135]. Thus, modelling competition is beyond the scope of the chapter. Furthermore, EVA is assumed as price-taker for energy and regulation market participation and submits only quantity demand/supply in them.

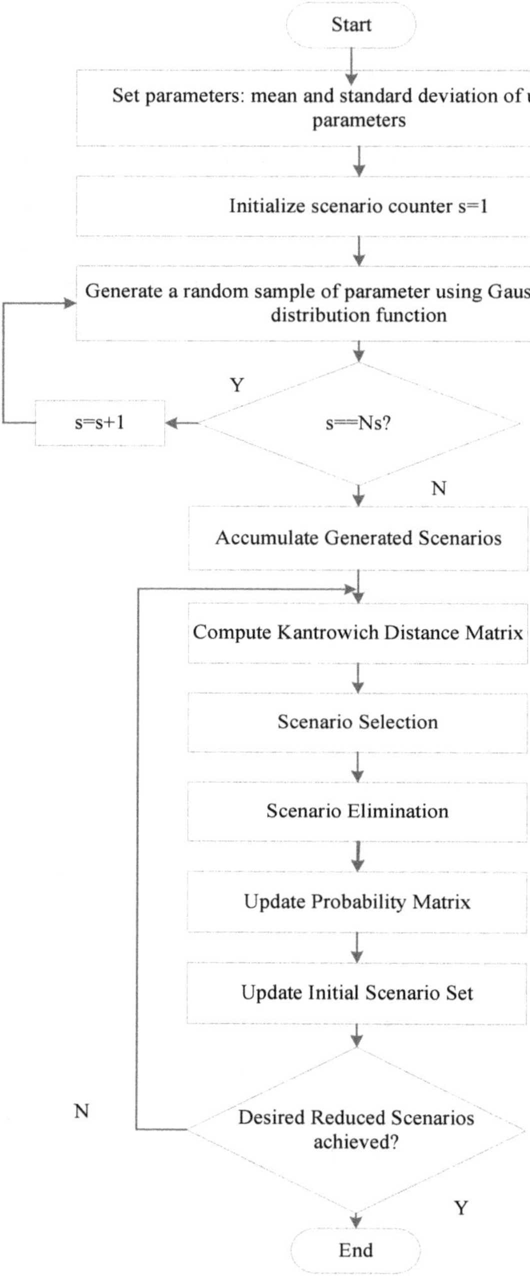

Fig. 4.4: Schematic Flowchart of Scenarios Generation and Reduction for Uncertain Variable

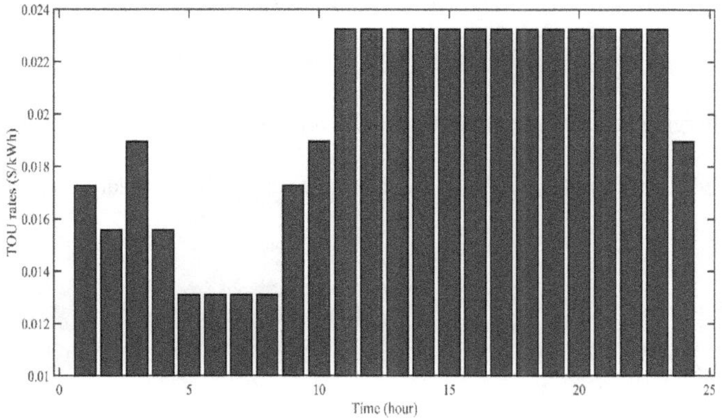

Fig. 4.5: Bar Plot of Five-Tiered TOU Pricing Incorporated in Charging Algorithm

4.5 Proposed Model for the System of Charging

The system model of EVs charging is the same as discussed in Chapter 3, comprising of a central controller (EVA) to perform optimal scheduling. Fig. 4.6 illustrates a graphical representation of the system, with additional factor for execution of PBDR as optimally designed TOU price. At the beginning of the day, SO communicates technical constraints and base loads to EVA. It is a precondition that the EVA is notified of the expected departure time whenever an EV is plugged in. At the beginning of the interval $t, t = t_1, t_2, \cdots, |T|$ where $|T|$ represents the total number of intervals in the set T (scheduling horizon), EVA communicates with EV owners to collect EV profile information.

EV profile information is composed of expected arrival time, departure time for the following day, and the initial SOC on entrance depending upon estimated strategic expeditions. It retrieves through EV ID the EV characteristic information like total battery capacity, permissible lower and upper limits of SOC of EV. Based on collected information, it determines the current ongoing EV set $N_{i,t}$ and the current charging period or available duration $|td_i - ta_i|$. Later, EVA plots the outline information for individual EV i to the corresponding characteristic information using EV ID.

The current charging-interval matrix is determined according to the number of EVs having similar arrival and departure pattern. In this scheme, the EVA receives the pricing information from the electricity market. Based on all available information, EVA decides optimal charging scheduling, satisfying both technical and economic objectives and finds the optimal charging rate $POP^*_{m,t}$ $(\forall m \in N_{i,t})$ by solving Dynamic Charge Scheduling Problem $(DCSP)$. EVA communicates EVs to this state, according to the charging schedule. In other words, EV $m(\forall m \in N_{i,t})$ is charged with the optimal scheduled charging rates $POP^*_{m,t}$ in the interval t, to provide ancillary services.

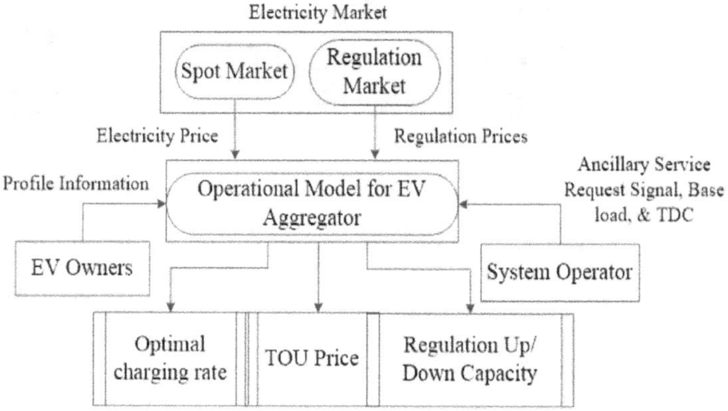

Fig. 4.6: Schematic Communication Flow of Proposed Integrated TOU-PBDR Dynamic G2V Scheduling

4.6 Problem Formulation

Fig. 4.7 details the simplified flowchart for the proposed optimization technique. The scheduling horizon for the problem considered is 24 hours for complete one day from 00 hours (midnight) to 23 hours on the same day. It is assumed that the commuter cars are plugged in and available for services between (8:00 to 17:00) and (22:00 to 6:00) providing sufficient time margin to charge EVs to the desired state while performing regulation services without comprising with the convenience of EV owners. EV owners have the flexibility of arrival and departure times. The initial SOC indicates the combined effects of battery loss due to distance travelled and other losses while travelling. Furthermore, scenarios for initial SOC falls in a narrow range (0.1-0.3) to consider the worst-case scenarios. Consequently, reflecting all possible losses due to mobility behaviour. The final departure/desired SOC was set to be 90 %.

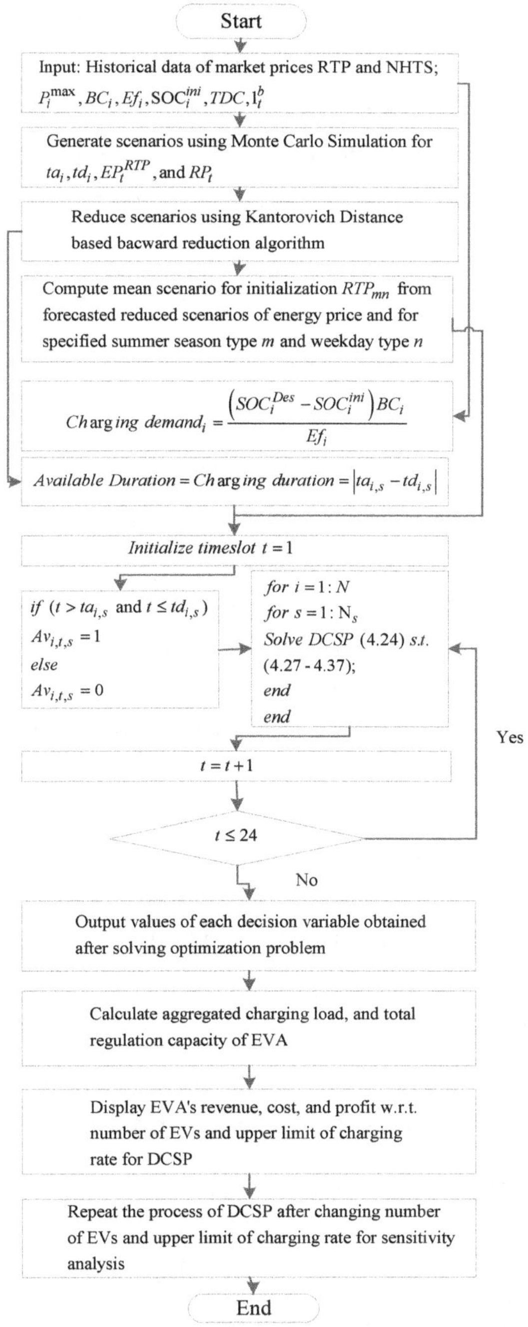

Fig. 4.7: Simplified Flowchart of Proposed Optimization Technique

Fig. 4.8 shows the real-time prices of the energy market and developed TOU rates for summer weekday in CAPITL zone of NYISO [134]. The profile of base-load is shown in Fig. 4.9 for the same zone. Analogous to Chapter 3, entrance time ta_i and leaving time td_i are forecasted from the historical driving form of the ith EV. It is presumed that they track normal pdf as in Eqs. (4.8) and (4.9):

$$pdf(ta_i) = \frac{1}{\sigma_{ta_i}\sqrt{2\pi}} \exp\left(-\frac{(ta_i - \mu_{ta_i})^2}{2(\sigma_{ta_i})^2} \right) \tag{4.8}$$

$$pdf(td_i) = \frac{1}{\sigma_{td_i}\sqrt{2\pi}} \exp\left(-\frac{(td_i - \mu_{td_i})^2}{2(\sigma_{td_i})^2} \right) \tag{4.9}$$

Here, (μ_{ta_i} and μ_{td_i}) are the mean (7:00, 19:00) and (σ_{ta_i} and σ_{td_i}) are standard deviation (2 hours) of the normally distributed ta_i and td_i respectively. The average charging window is from 19:00 to 07:00 since mass EVs are available for charging at this time. Feasibility of resultant schedules of charging would be validated when loading demand of system and EV owners' requirements are fulfilled as well. Imitations have been executed for thousand EVs for a characteristic EV user who frequently trips among residence and office. It is challenging to decrease peak implicitly. Thus, charge scheduling of EVs requires coordination with SO and EV owners both, necessitating at least 10 hours for slow charging. These EVs must be plugged in not delayed than 21:00 hour (19+2hours) to achieve the desired level of charging at departure 7:00 hour the subsequent morning. A smooth load profile could be attained by incorporating flexibility constraint (4.10), ensuring support to grid stability.

$$(SOC_{i,t} - SOC_{i,t-1}) \leq \frac{f_k \cdot \left(SOC_i^{DES} - SOC_i^{INI}\right)}{(ta_i - td_i + 1)} \tag{4.10}$$

Moreover, EVA is required to fulfil the charging requirement of EVs. Therefore, at the departure hour, the charging rate of an EV i is taken as a constraint (4.11),

$$POP_{i,t} = \left(SOC_i^{Des} - SOC_{i,t-1}\right) \cdot \frac{BC_i}{Ef_i} \cdot Av_{i,t} \quad \forall t \in td_i \tag{4.11}$$

The following assumptions are made in the scheduling model of charging:

i. SO sends dispatching signals to EVA, it does not schedule the EVs directly.

ii. An EV will be scheduled by EVA as soon as it arrives at EVA.

iii. EV can only charge when plugged in.

iv. EVA must guarantee driving requirements such that the battery is sufficiently charged for the next possible trip. EVs are charged to their desired SOC level by departure time.

v. New entrances and leavings of EVs happen on the finish of the specific time interval, but not within it. Thus, the charging power for the complete interval is the same.

vi. EV charging is never stopped until it is charged to the desired SOC once it is scheduled.

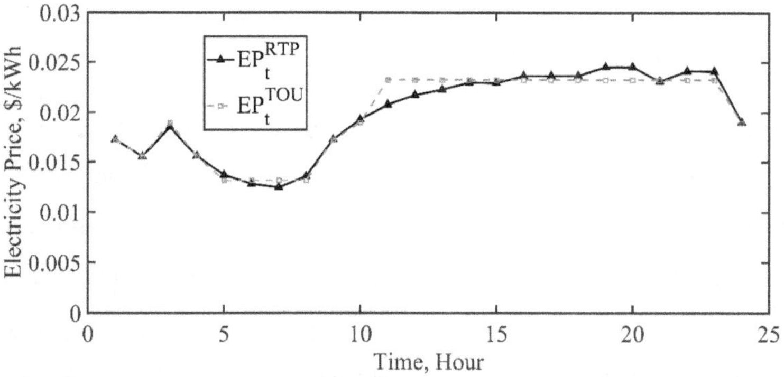

Fig. 4.8: Developed five-tiered TOU Rates to Reflect Wholesale Electricity Real-Time Prices for a Summer Weekday

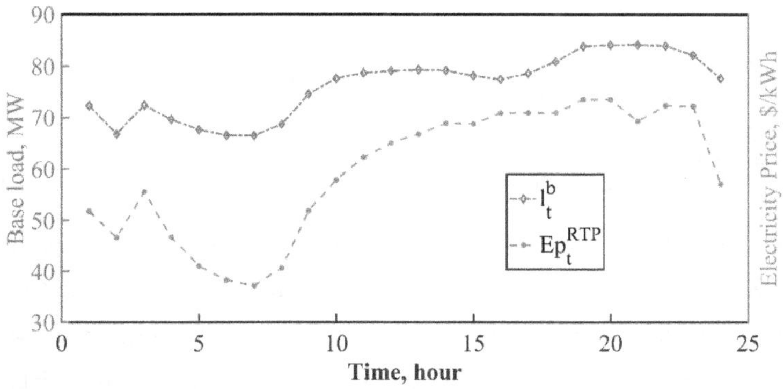

Fig. 4.9: Base Load Profile and Real-Time Price for a Summer Weekday

As discussed in Chapter 3, unregulated/uncontrolled charging is the baseline type of charging, in which the EVA charges an EV in a single time slot as soon as it is plugged into the grid. Therefore, such charging is non-price responsive. In this uncontrolled charging or baseline setup, EVs are charged by the maximum rate of charging $POP_{i,t}^{\max}$ as expressed by Eq. (4.12).

$$POP_{i,t}^{\max} = \left(SOC_{i,t}^{DES} - SOC_{i,t}\right) \cdot BC_i / E_{f_i} \forall i,t \tag{4.12}$$

$Av_{i,t}$ being the binary variable gives the availability of EV i which is plugged in during individual time interval t as per Eq. (4.13). The total charging energy cost of unregulated charging may represent cost upper bound. This cost bound (CB) is the higher value of charging cost corresponding to the availability of EVs as Eq. (4.14).

$$Av_{i,t} = \begin{cases} 1 & ta_i \leq t \leq td_i \\ 0 & otherwise \end{cases} \tag{4.13}$$

$$CB = \sum_{t=1}^{T} \sum_{i=1}^{N} POP_{i,t}^{\max} \cdot \left(M_t + EP_t^{RTP}\right) \cdot Av_{i,t} \tag{4.14}$$

Aforementioned in Chapter3, for the static charging scenario, EVA has complete knowledge of charging profile information of ith EV: ta_i, td_i, SOC_i^{INI}, and SOC_i^{DES}. In the present work, EVA procures energy from the wholesale market at a real-time price $\left(EP_t^{RTP}\right)$ and sells it at a 40% mark-up (M_t) on designed TOU rates $\left(EP_t^{TOU}\right)$. This work formulates SCSP as a profit maximization problem of EVA as a difference in its revenue and costs to quantify the dynamics of the problem. Therefore, the scheduling problem can be defined via the statement (4.15):

$$\max_{SOC_{i,t}, POP_{i,t}, rc_{i,t}} \Pr ofit = \operatorname{Re} v^{SCSP} - CC^{SCSP} \tag{4.15}$$

$$\operatorname{Re} v^{SCSP} = \sum_{t=1}^{T} \sum_{i=1}^{N} RP_t \cdot rc_{i,t} + \sum_{t=1}^{T} \sum_{i=1}^{N} \left(M_t + EP_t^{TOU}\right) \cdot POP_{i,t} \cdot Av_{i,t} \tag{4.16}$$

$$CC^{SCSP} = \sum_{t=1}^{T} EP_t^{RTP} \left(\sum_{i=1}^{N} POP_{i,t} \cdot Av_{i,t} \right) \tag{4.17}$$

The revenue of EVA is shown in (4.16) with two terms explained in Chapter 3. For the specified charging tasks set, the revenue of EVA must be highest amongst all viable tasks of charging in contradiction to the purchase cost (CC^{SCSP}) which as per (4.17) should be minimum. Feasible charging task satisfies both economic constraints of EV and technical constraints of the system. SCSP (4.15) is subject to constraints (4.18-

4.24), whose explanation is already discussed in Chapter 3. The overall regulation capacity is equivalent to the sum of the regulation up and regulation down capacities as Eq. (4.21). Eqs (4.22) and (4.23) respectively illustrate how EV capacity can be used for providing regulation down and regulation up capacity. Congestion management constraint is expressed as Eq. (4. 24).

$$SOC_{i,t} = \begin{cases} SOC_i^{INI} & \forall t = \lfloor ta_i \rfloor \\ SOC_i^{DES} & \forall t = \lceil td_i \rceil \\ SOC_{i,t-1} + Ef_i Av_{i,t-1} POP_{i,t-1} / BC_i & ,otherwise \end{cases} \tag{4.18}$$

$$P_i^{\min} \leq POP_{i,t} \leq P_i^{\max} \forall i,t \tag{4.19}$$

$$\sum_{t=1}^{T} \left(M_t + EP_t^{TOU} \right) \sum_{i=1}^{N} \left(POP_{i,t} \cdot Av_{i,t} \right) \leq CB \tag{4.20}$$

$$rc_{i,t} = uc_{i,t} + dc_{i,t} \tag{4.21}$$

$$uc_{i,t} = \left(POP_{i,t} - P_i^{\min} \right) \cdot Av_{i,t} \tag{4.22}$$

$$dc_{i,t} = \left(\min\left(POP_{i,t}^{\max}, P_i^{\max} \right) - POP_{i,t} \right) \cdot Av_{i,t} \tag{4.23}$$

$$\sum_{i=1}^{N} (POP_{i,t} + dc_{i,t}) + l_t^b \leq TDC \tag{4.24}$$

4.6.1 Dynamic Charging Scheduling Problem (DCSP)

As discussed earlier, in DCSP, EVA doesn't have prior knowledge of the charging assignments. Whenever an EV is plugged into a charging network on arrival, the problem is solved to determine its charging schedule. The proposed DCSP problem is of much-reduced dimensions with constraints and variables belonging to a specific single i^{th} EV at a specific time t as compared to the conventional static problem considering all EVs at that time. Due to lack of information regarding charging profile, EVA characterizes mobility uncertainty aforementioned in section 3. For stochastic DCSP considering mobility uncertainty scenario s with probability π_s, EVA's profit can be defined as Eq. (4.25):

$$\max_{SOC_{i,t,s}, POP_{i,t,s}, rc_{i,t,s}} \text{Profit}_{DCSP} = \sum_{s=1}^{N_s} \pi_s \cdot \text{Profit}_s$$

$$\text{Where, Profit}_s = \left(A\text{Rev}_{DCSP}(s) - ACC_{DCSP}(s) \right) \tag{4.25}$$

Subject to constraints (4.27-4.37).

$$\underset{SOC_{i,t,s},POP_{i,t,s},rc_{i,t,s}}{A\mathrm{Re}v_{DCSP}(s)} = \sum_{i=1}^{N} \mathrm{Re}v_{DCSP}(i,s) \quad ;$$

$$\underset{SOC_{i,t,s},POP_{i,t,s},rc_{i,t,s}}{ACC_{DCSP}(s)} = \sum_{i=1}^{N} CC_{DCSP}(i,s)$$

$Rev_{DCSP}(i,s)$ and $CC_{DCSP}(i,s)$ can be solved independently or simultaneously. Such independent optimization of revenue and charging cost may give locally optimal solutions. Therefore, to achieve optimal aggregated revenues to EVA ($ARev_{DCSP}(s)$) and costs ($ACC_{DCSP}(s)$), simultaneous optimization is performed. Thus, the mathematical formulation of revenue to EVA from i^{th} EV in s^{th} scenario of DCSP ($Rev_{DCSP}(i,s)$) can be specified as (4.26):

$$\underset{SOC_{i,t,s},POP_{i,t,s},rc_{i,t,s}}{Rev_{DCSP}(i,s)} = \sum_{t=1}^{T} RP_t \cdot rc_{i,t,s} + \sum_{t=1}^{T} \left(M_k + EP_t^{TOU}\right) \cdot POP_{i,t,s} \cdot Av_{i,t,s} \tag{4.26}$$

Revenue from EV i is similar to SCSP consisting of two terms as regulation revenue and charging energy revenue, but its value is different according to different mobility scenarios s $\left(t_{a_{i,s}}, t_{d_{i,s}}, SOC_{i,s}^{INI}\right)$. For DCSP, Eq. (4.27) illustrates the SOC of EV i in a time interval t scenario s, Eq. (4.28) shows limits of charging rate, and (4.29) is the cost bound. Constraints (4.30-4.32) are total regulation capacities, up and down capacities respectively. The constraints and objective function in this charging scheduling problem formulation are for recently incoming i^{th} EV. Remarkably EVA takes into account this recent i^{th} EV and all those EVs that were plugged in before i^{th} EV in Eq. (4.33) as congestion management constraint. Constraint (4.34) is the flexibility limitation controlling change in SOC between two consecutive intervals from available duration and charging rate. Equation (4.35) is the maximum charging rate defined as charging to desired SOC from the current SOC in one timeslot. (4.36) indicates deadline constraint to complete charging by departure time. If $t \in \left[t_{a_{i,s}}, t_{d_{i,s}}\right]$, (4.37) illustrates the availability of EV i in the time interval t for the scenario s.

$$SOC_{i,t,s} = \begin{cases} SOC_{i,s}^{INI} & \forall t = \lfloor ta_{i,s} \rfloor \\ SOC_{i,s}^{DES} & \forall t = \lceil td_{i,s} \rceil \\ SOC_{i,t-1,s} + Ef_i Av_{i,t-1,s} POP_{i,t-1,s} / BC_i & , otherwise \end{cases} \tag{4.27}$$

$$P_{i_{\min}} \leq POP_{i,t,s} \leq P_{i_{\max}} \; \forall t, \forall s \tag{4.28}$$

$$\sum_{t=1}^{T} \left(M_k + EP_t^{TOU} \right) POP_{i,t,s} Av_{i,t,s} \leq CB \tag{4.29}$$

$$rc_{i,t,s} = uc_{i,t,s} + dc_{i,t,s} \quad \forall t, \forall s \tag{4.30}$$

$$uc_{i,t,s} = \left(POP_{i,t,s} - P_{i_{\min}} \right) \cdot Av_{i,t,s} \quad \forall t, \forall s \tag{4.31}$$

$$dc_{i,t,s} = \left(\min\left(POP_{i,t,s}^{\max}, P_{i_{\max}} \right) - POP_{i,t,s} \right) \cdot Av_{i,t,s} \quad \forall t, \forall s \tag{4.32}$$

$$\sum_{i=1}^{m} POP_{i,t,s} + l_t^b + \sum_{i=1}^{m} dc_{i,t,s} \leq TDC \quad \forall t, \forall s \tag{4.33}$$

$$(SOC_{i,t,s} - SOC_{i,t-1,s}) \leq \frac{f_k \cdot \left(SOC_i^{DES} - SOC_{i,s}^{INI} \right)}{(ta_{i,s} - td_{i,s} + 1)} \tag{4.34}$$

$$POP_{i,t,s}^{\max} = \left(SOC_i^{DES} - SOC_{i,t,s} \right) \cdot BC_i / E_{f_i} \forall i, t, s \tag{4.35}$$

$$POP_{i,t,s} = \left(SOC_i^{Des} - SOC_{i,t,s} \right) \cdot \frac{BC_i}{Ef_i} \cdot Av_{i,t,s} \; \forall \mathrm{t} \in \mathrm{td}_{i,s} \tag{4.36}$$

$$Av_{i,t,s} = \begin{cases} 1 & t_{a_{i,s}} \leq t \leq t_{d_{i,s}} \\ 0 & otherwise \end{cases} \tag{4.37}$$

In the same way, the prime problem of the EVA's purchase cost due to i^{th} EV in reduced s^{th} scenario ($CC_{DCSP}(i,s)$) can be expressed by the following objective function (4.38):

$$\underset{SOC_{i,t,s}, \mathrm{POP}_{i,t,s}, rc_{i,t,s}}{CC_{DCSP}(i,s)} = \sum_{t=1}^{T} \left(M_t + EP_t^{RTP} \right) \cdot POP_{i,t,s} \cdot Av_{i,t,s} \tag{4.38}$$

Subject to constraints: (4.27-4.37).

Dynamic charging scheduling with updating the charging rate of each EV according to both prices (4.29) and load (4.33) constraints improves its performance. It modifies the charging rates for plugged-in available EVs every single time interval, 'm' additional EVs reach responding to TOU prices and baseload from SO. This updating DCSP is achieved by similarly solving the problem to the static scheduling problems $\mathrm{Re}\,v^{SCSP}$ or CC^{SCSP}, excepting that input, comprises EVs that are plugged in at this time to the system (as a replacement for all EVs). For the lower value of m, performance is improved regarding economic efficiency at the trade-off of higher computational time from stochasticity. The objective function and the constraints for articulating the scheduling problem have been formulated referring to [53, 65, and 136]. An optimization framework is created in General Algebraic Modelling System (GAMS) [137]. Exploring the characteristics of constraints and objective functions, it is perceived that whole optimization is Mixed-Integer Non-Linear Problem (MINLP) with endogenous parameters, and therefore, is solved using the COIN-OR IPOPT (Interior Point Optimizer) solver suitable for large-scale MINLPs in GAMS, and the outcomes are described comprehensively in section 4.7.

4.7 Results and Discussions

To confirm the scalability of the proposed approach, 1000 EVs representing 30% penetration have been accommodated. Illustrating travelling behaviour was found from the NHTS database [93]. The impact of pragmatic aggregated charging on the system is analyzed with sensitivity analysis. To validate the results of optimization, the average load profile for the summer season has been considered from [134].

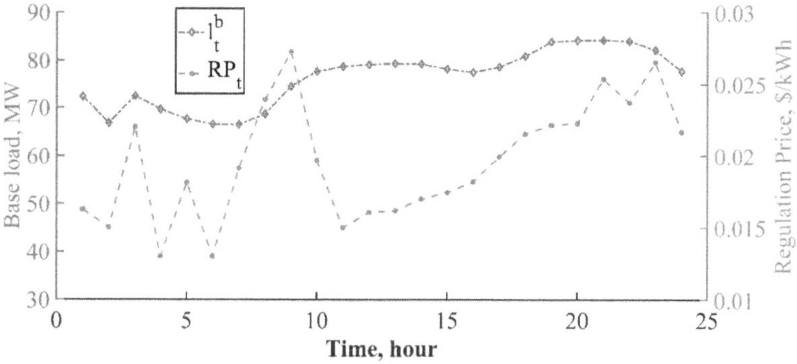

Fig. 4.10: Prices for Regulation Capacity in $/kWh and Baseload Profile

Particularly, for each day, the regulation capacity prices and hourly electricity prices were obtained from the same. Five-tiered TOU prices with off-peak, shoulder-off peak, flat rate, shoulder-peak, and peak heights are developed employing the AHC method. Regulation prices are as shown in Fig.4.10. As observed in Fig. 4.10, for off-peak load hours, regulation prices are higher as SO calls for regulation down capacity for these hours. Regulation services are provided to SO when EVs are available (t_1-t_6 and t_{19}-t_{24}), and prices are higher, as shown in Fig. 4.11. Even though regulation prices are high, when most of the EVs are not available (t_7-t_{18}), significant regulation services are not provided. As compared to charging revenue, regulation revenue is high, since short-term regulation service prices are more volatile.

Baseload profile is in unison with energy prices. The charging load is also harmonized according to the TOU-PBDR of EVs. When prices are lower, the scheduled charging rate is higher and vice-versa. Therefore, the aggregated scheduled charging rate is showing Pareto-optimal behaviour with baseload/ prices of energy. Fig. 4.12 shows that due to the proposed coordinated approach, for lower values of baseload, aggregated charging load is higher due to higher scheduled charging rates, i.e., the EVs are charged in off-peak duration. Charging load reduction during peak hours assists in improving the load factor of the system. Incorporating frequency control through regulation services, the aggregated EVs are managing peak load demand by keeping lower charging rates for peak duration and thereby support the grid stability and reliability enhancement. Besides, modulation of charging rates resulting in controlled charging behaviour alleviates charging load pressure on, and overload risk of, the transformer.

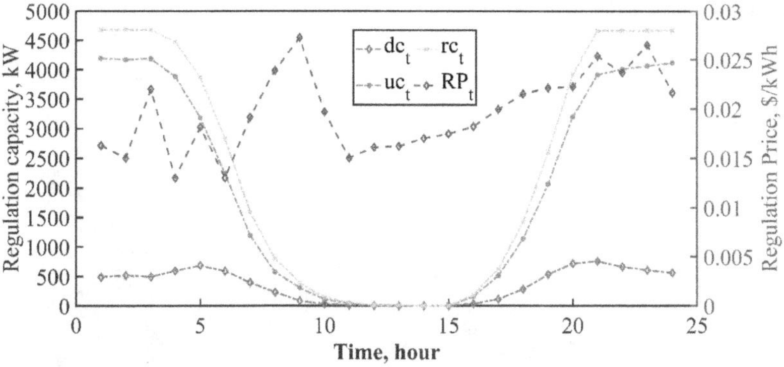

Fig. 4.11: Regulation Capacity Provided by EVA at a Time *t* in Response to Regulation Price Signals

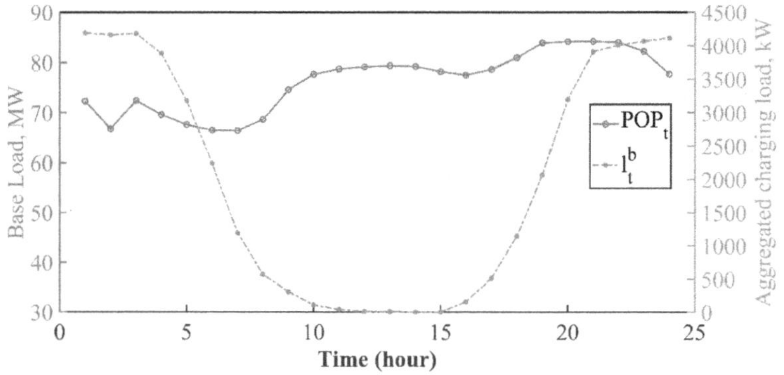

Fig. 4.12: Aggregated Scheduled Charging Rate Adaptation in Response to Baseload Profile

Unregulated charging has the unfavourable consequence of triggering an upsurge in peak load on the system. Unregulated charging cost gives the upper bound of charging cost (CB). Figs. 4.13 and 4.14 depict sensitivity analysis of EVA's performance metrics with charging cost minimization and profit maximization objectives respectively under various charging methods for varying number of EVs. It is observed that purchase cost, revenue to EVA and charging cost to EV owners increases with the number of EVs. As compared to unregulated charging, the cost is lower in dynamic coordinated charging with minimum under SCSP.

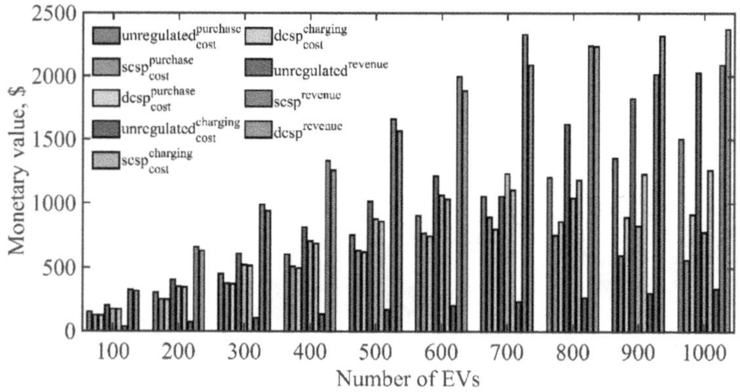

Fig. 4.13: Sensitivity Analysis of EVA Performance Metrics with Charging Cost Minimization as Objective

Irrespective of objective or perspective concerned, regulation revenue earned are higher as compared to charging revenue. Therefore, the revenue of EVA increases with the significant regulation capacity for a large number of EVs. The sensitivity of revenue and purchase cost is lower as compared to charging cost to EV owners, for a varying number of EVs. The price-responsive behaviour of EV owners motivates them to remain connected to EVA. Initially, for a smaller group of EVs, DCSP gives lower revenue as compared to SCSP, at the trade-off of decreased profits. Conversely, for a larger group of EVs, EVA earns higher revenue as compared to SCSP, at the trade-off of increased profits due to reduced costs further.

From Fig. 4.15, it is clear that SCSP performs better than DCSP for a smaller group of EVs, but performance degrades as compared to DCSP after a 21 % penetration point. Since EVA incurs a lower profit in SCSP after N=700. SCSP will be beneficial up to this point, but with an increase in the number of EVs, profit reduces which is undesirable. DCSP is guaranteeing profit without restricting on number of EVs. Since EVA can hedge the risk of any loss in the case of DCSP, it gives beneficial results from the perspectives of all entities. For validation, Table 4.1 indicates purchase cost and revenue for 1000 EVs for unregulated and coordinated scheduling strategies.

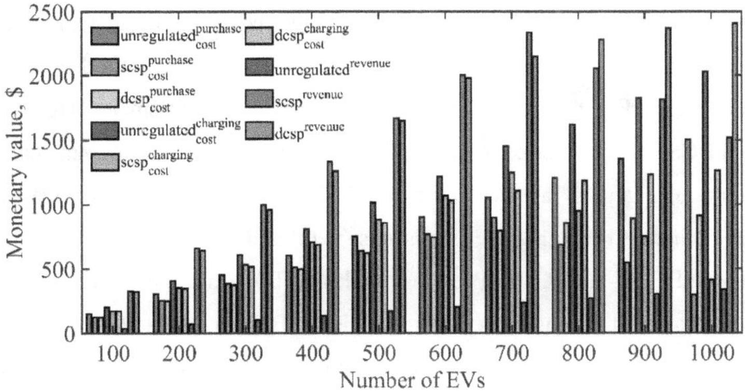

Fig. 4.14: Sensitivity Analysis of EVA Performance Metrics with Profit Maximization as Objective

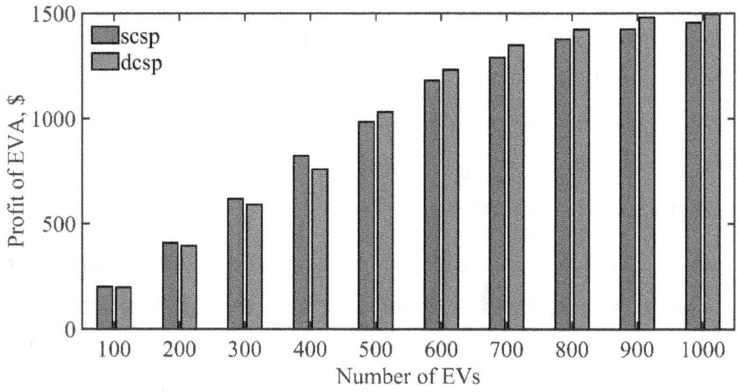

Fig. 4.15: Profit of EVA in Integrated TOU-PBDR and SCSP/ DCSP w.r.t. the Number of EVs

Fig. 4.16 shows the sensitivity analysis of purchase cost, charging cost and revenue earned by EVA for varying upper limit of charging rate and different objectives (P- illustrates profit maximization whereas C- depicts cost minimization objectives). It is clear from Fig.4.16 that sensitivity of revenue is lower in TOU-PBDR integrated DCSP while that of the purchase cost and charging cost is higher with the adaptation of the upper charging rate limit. Hence, profit sensitivity is also lower being a difference in revenue and cost. Fig. 4.17 shows the charging load on the system in TOU-PBDR integrated DCSP of EVA for coordinating group of 1000 EVs, with different adaptive upper limits of charging rate, *i.e.* 6.6 and 9.9. The charging load is considered as aggregated scheduled charging rates in response to offered TOU prices. Valley is filled with proposed TOU-PBDR DCSP during the lower TOU rates period (t_{22}-t_3)/ night hours, and peak load management during the higher TOU rates span (t_{11}-t_{15}). For the higher value of the upper charging rate limit, smooth variation in charging rates of an EV between two consecutive time intervals is hampered. Therefore, a higher limit of the upper charging rate would ultimately lead to uneven charging load and hence threat to grid stability.

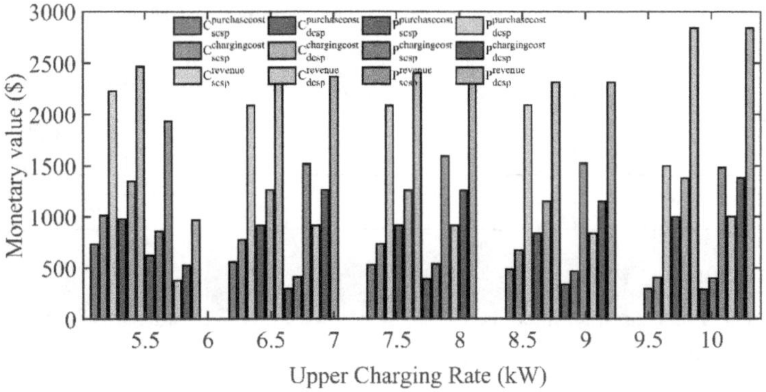

Fig. 4.16: Sensitivity Analysis of Purchase cost, Charging cost and Revenue for Varying Upper Charging Rate Limit

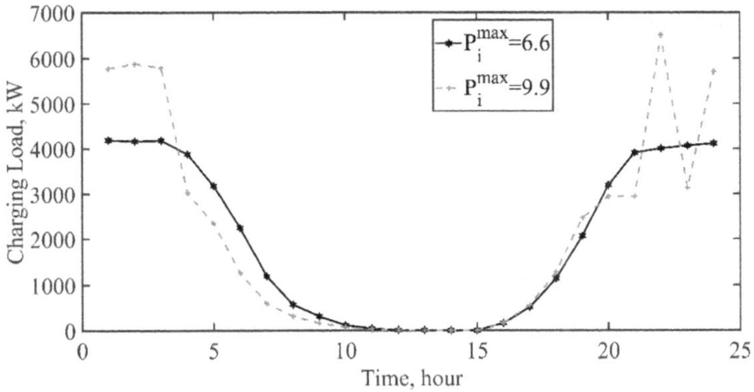

Fig. 4.17: Charging Load in TOU-PBDR Integrated DCSP for 1000 EVs with Different Upper Charging Rate Limits

An individual optimum for EVA/ EV owner does not necessarily indicate a system optimum. There exists Pareto-optimality among economic and technical objectives. From the results of static and dynamic charge scheduling, dynamic charging algorithm outshone static algorithm by giving much higher profits, for a larger group of EVs. The proposed business model of EVA is beneficial to EVA, SO and EV owners promoting EV proliferation and green transportation in sustainable energy and smart grid environment. The minimized charging costs and required SOC level for EV owner are the motivation for EVs participation in grid service provision. The proposed day-ahead strategy enhances both the economic efficiency of SO and cost-effective modelling of the price-responsive behaviour of EV owners.

Table 4.1 Purchase Cost and Revenue (Without and With Coordination)

	Unregulated Charge Scheduling	SCSP	DCSP
Purchase cost ($)	1504.032	293.4196	913.1859
Revenue ($)	332.9503	1517.61	2368.294

4.8 Epilogue

In this chapter, an integrated TOU-PBDR and dynamic G2V charge scheduling of EVA is modelled to optimize regulation services and charging cost simultaneously. TOU-PBDR avoids system overloads throughout the scheduling horizon and hence defers distribution system upgrades. In the proposed scheme, the charging rate is evaluated for each EV and a scenario based on baseload profile and TOU prices. A comparative analysis of integrated TOU-PBDR and G2V DCSP is done with SCSP and unregulated charging schemes. It shows that the proposed scheme under optimal TOU rate structure gives the optimal results as compared to unregulated charging. Due to more flexibility from EV owners, EVA earns more profit in DCSP as compared to SCSP for larger groups. This increase in profit is due to the optimal price-responsive behaviour of EV owners in DCSP. Sensitivity analysis of EVA performance metrics indicates that DCSP is more profitable as compared to SCSP with an increase in the number of EVs although cost increases but revenue also increases to enhance the overall profit. Moreover, TOU-PBDR integrated DCSP approach also helps in attaining equilibrium between EVA and EV owners' financial concerns. The proposed approach provides prior information of next day load and would assist SO in regulation product bidding.

Further scope of this work could be an investigation of the price uncertainty impact on optimal regulation offering strategies. i.e. Risk-based dynamic G2V scheduling with the integration of TOU-PBDR and considering multiple uncertainties would be discussed in the subsequent chapter.

Chapter 5

RISK-BASED INTEGRATED DR AND STATIC G2V SCHEDULING OF EVA FOR MULTIPLE MARKETS

The work anticipated in Chapter 3 can be further extended by incorporating demand response and risk management of EVA. The DR integration reduces EV owners' charging costs in the electricity market, and it also considers techno-economic constraints of SO and EV owners, promising the excellence of V2G services by EVA. A robust risk management plan has the potential to help an EVA establish procedures to avoid potential threats, minimize their impact should they occur and cope with the results. This ability to understand and control risk enables EVA to be more efficient in its decision-making. Furthermore, focusing on risk management can help an EVA reach its goals. This chapter proposes a risk-based integrated DR and static G2V scheduling of EVA for multiple market products considering grid stability concern.

The chapter is organized as follows. An introduction highlighting the significance of the risk-based G2V scheduling model followed by a brief overview of the existing work is provided in Section 5.1. In Section 5.2, risk controlling in stochastic optimization is illustrated. The techniques for scenarios generation and reduction are represented in Section 5.3. The risk-based stochastic programming problem (SPP) formulation is detailed out in Section 5.4. Subsequent Section 5.5 discusses the results obtained through simulations of risk-based stochastic scheduling. Finally, Section 5.6 concludes the outcomes of this chapter.

5.1 Introduction

EVA acts as a mutually beneficial negotiator between SO & EV owners, enabling improved operations in the power grid and electricity market. It aims to optimally schedule the charging of EVs considering the benefits of SO and EV owner while maximizing its profit earned from participation in multiple markets of electricity and ancillary services. It provides regulation capacity and grid congestion management through the coordinated charging of EVs for the benefit of SO. It may adopt PBDR to benefit EV owners by reduced charging costs and benefit SO by peak load management and load levelling. As discussed in chapter 4, TOU-PBDR shields EV owners against price volatility. However, for optimal decision making, EVA is subjected to risk from volatility in different market prices. Therefore, the incorporation of a risk model based on different uncertainties is necessary for realistic decision making in the G2V scheduling.

The price uncertainty would significantly affect market operations and behaviour of an EVA aimed at profit maximization through G2V charge scheduling of EVs in a smart grid environment. In this context, for improved multi-market operations, this chapter proposes risk-based G2V scheduling of EVA in energy and regulation markets, to manage its revenue risk and reduce energy purchase cost for EV owners incorporating TOU-PBDR. Five-step TOU price is formulated with the forecasted value of real-time energy prices using the hierarchical clustering method. The energy and regulation market prices are modelled by a normal distribution with forecasted as mean and a standard deviation to model uncertainty pattern. Scenarios of prices are generated employing the Monte Carlo Simulation approach. The probabilistic Kantorovich Distance (KD) is used in scenario reduction backward algorithm to find a trade-off between computation speed and accuracy. EVA procures energy from the wholesale market on real-time market price and charges EVs on TOU prices. The EVA also provides regulation to SO at the real-time regulation market prices by coordinating the charging rates of connected EVs. The objective of the EVA is expected profit maximization subjected to risk of profit variation due to volatility of the uncertain variables like market prices of energy and regulation market. The risk measure incorporated is Conditional Value at Risk ($CVaR$). Simulation results verify the efficacy of the proposed method and the effect of the stochastic G2V scheduling strategy of EVA on the energy and regulation market operations.

5.2 Risk Controlling in Stochastic Optimization

Risk control is an evaluation of an EVA's willingness and ability to take risks. It can also refer to the threats to which an EVA is exposed. Risk control is important for managing risk cost-effectively by determining a proper investment asset allocation for a portfolio. EVA can use risk control as a way to mitigate potential risks and threats. In line with this, the risk appetite of EVA represents the amount and type of risk that an

EVA is willing to take to meet its strategic objectives. Risk appetite is the total aggregated risk that the EVA can bear in a given risk control as compared to risk tolerance indicating the level of risk that an EVA can accept per individual risk.

Risk consists of three parts: an uncertain situation, the likelihood of occurrence of the situation, and the effect (positive or negative) that the occurrence would have on the performance of EVA's G2V scheduling. Risk assessments are very important as they form an integral part of any management strategy. They help to: Create awareness of hazards and risk. Identify who may be at risk (e.g., EVA, EV owners, SO, etc.). The purpose of risk management is to identify potential problems before they occur, or, in the case of opportunities, to try to leverage them to cause them to occur. Risks management empowers EVA's business model with the necessary tools so that it can adequately identify potential risks.

To govern the risk of experiencing low profit with a high probability, risk control constitutes an important issue in stochastic programming problem (SPP) formulation. Therefore, an additional term measuring the risk associated with a profit distribution is incorporated in the SPP formulation for risk management. The risk-averse decision-making problem ($RADM$) is expressed as

$$RADM = (1-\beta)\cdot Expected\ \Pr ofit + \beta \cdot Risk\ measure$$

Thus, if β = 0, the risk measure is neglected, while the expected profit is disregarded if β = 1.

The Decision-making framework of EVA is challenged by market prices. Due to their volatility and uncertainties, EVA's expected profit varies from the mean value. As demonstrated in Fig. 5.1, Value at Risk (VaR) represents the largest value η ensuring the probability of obtaining profit less than η is lower than $1-\alpha$. α is the confidence interval. However, VaR does not detect fat tail in profit distribution. Tail risk in EVA's portfolio is quantified by a coherence risk measure termed as $CVaR$. It is the expected value of profit smaller than $1-\alpha$ quantile of profit distribution, imitative of the weighted average of "worst-case" losses in the tail of the probability distribution of expected profits, beyond VaR.

$CVaR$ is the expected value of VaR, which can be mathematically interpreted as Eq. (5.1).

$$CVaR_{\alpha,s} = \max\left\{\eta - \frac{1}{1-\alpha}\pi_s\left\{\max\left\{\eta - \text{Profit}_s, 0\right\}\right\}\right\}, \forall\alpha \in (0,1) \quad (5.1)$$

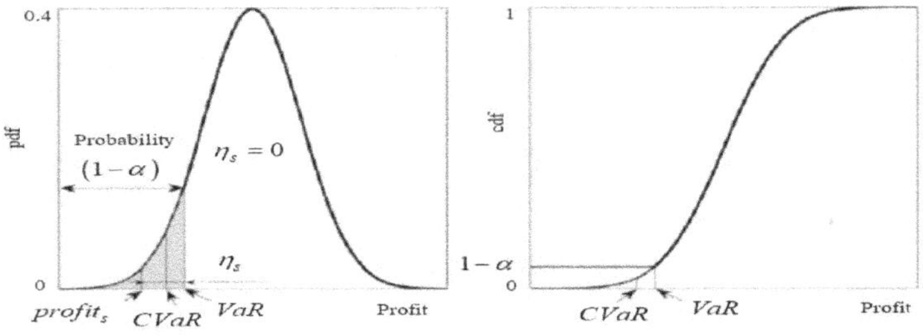

Fig. 5.1: Demonstration of VaR *and* $CVaR$ **concepts**

5.3 Scenario Generation and Reduction

As discussed in Chapter 4 and Section 4.3, precise uncertainty depiction imposes to consider a huge number of scenarios, therefore to deal with associated computational complexity, the inconsistent method for scenario-reduction is necessitated. This chapter adopts the method discussed in Section 4.3 for scenario generation and reduction of energy and regulation prices. One thousand scenarios of prices are generated employing the Monte Carlo Simulation approach. The generated scenarios are reduced to ten, utilizing the KD based backward method as presented in [126] from MATLAB simulations. The KD matrix involves the multiplication of price function and matching incidence likelihood π_s of all produced scenarios as depicted in Eq. (5.2). It is used for producing decreased scenario sets that are near enough to the initial set. The cost matrix comprises the price functions $\upsilon(s,s')$ for overall produced sets of scenarios.

$$KD(s,s') = \sum_{s\in\Omega/\Omega_{subset}} \pi_s \cdot \upsilon(s,s') \quad (5.2)$$

The price function is described as a vector difference between scenario subsets s & s', and calculated using (5.3).

$$\upsilon(s,s') = \left\|C_s - C_{s'}\right\| \quad \forall s,s' \in \Omega, \Omega = 1,2,\cdots,N_s \quad (5.3)$$

5.4 Risk-Aversive Formulation of Stochastic Programming Problem

It is assumed that arrival, departure times and energy usage are known to EVA. *CVaR* risk measure is employed and added as constraints. Risk-averse and risk-neutral G2V charge scheduling are considered with impact analysis of risk-aversion degree β

in energy and regulation markets. The risk-averse decision-making objective function of EVA is Eq. (5.4). Constraints (5.5-5.7) models risk measure to control the variability of expected profit. Risk measure $CVaR$ is defined as (5.5).

$$\max_{POP_{s,i,t}, rc_{s,t}, s_{\omega}, \eta} (1-\beta)\sum_{s=1}^{N_s}\sum_{i=1}^{N}\sum_{t=1}^{T}\pi_s \cdot RP_{s,t}^{sell} \cdot rc_{s,i,t} + EP_t^{TOU} \cdot POP_{s,i,t} - \pi_s \cdot EP_{s,t}^{RTP} \cdot POP_{s,i,t} + \beta \cdot \left(\eta - \frac{1}{1-\alpha}\sum_{s=1}^{N_s}\pi_s \cdot \phi_s\right) \tag{5.4}$$

$$CVaR = \eta - \frac{1}{1-\alpha}\sum_{s \in N_s}\pi_s \cdot \phi_s \tag{5.5}$$

$$\eta - \left(\sum_{i=1}^{N}\sum_{t=1}^{T} RP_{s,t} \cdot rc_{s,i,t} + EP_t^{TOU} \cdot POP_{s,i,t} - EP_{s,t}^{RTP} \cdot POP_{s,i,t}\right) \le \phi_s, \forall s \in N_s \tag{5.6}$$

$$\phi_s \ge 0 \tag{5.7}$$

The scheduled charging rate $POP_{s,i,t}$ is the value of power consumed by the EV owner. For risk-aversive problem, ϕ_s is a continuous and non-negative auxiliary variable, is equal to the maximum value between $\eta - \left(\sum_{i=1}^{N}\sum_{t=1}^{T} RP_{s,t} \cdot rc_{s,i,t} + EP_t^{TOU} \cdot POP_{s,i,t} - EP_{s,t}^{RTP} \cdot POP_{s,i,t}\right)$ and 0. EP_t^{TOU} are the predetermined and constant sell prices designed to reflect day-ahead wholesale prices. Constraint (5.8) indicates the updating of SOC of ith EV at time t. Rated capacity is BC_i (kWh), and the charging efficiency of ith EV battery is Ef_i. Constraint (5.9) shows the flexibility limitation for smooth modulation of the charging rate. Equation (5.10) provides the limits of charging rates. The capacity limit of the battery charger for i^{th} EV is $P_i^{\max}$. A binary variable representing the availability of EV $Av_{i,t}$ indicates whether EV is plugged in or not. Its value is 1 for EV's availability. Equation (5.11) ensures that EVs are charged only when connected and not at the time of travel. Constraint (5.12) assures it desired SOC at departure to EV owners.

$$SOC_{s,i,t} = SOC_{s,i,t-1} + \frac{Ef_i \cdot POP_{s,i,t}}{BC_i} \forall s, \forall i, \forall t \tag{5.8}$$

$$\left(SOC_{s,i,t} - SOC_{s,i,t-1}\right) \le f_k \cdot \frac{SOC_i^{Des} - SOC_i^{INI}}{t_{d_i} - t_{a_i}} \forall s, \forall i, \forall t \tag{5.9}$$

$$P_i^{\min} \le POP_{s,i,t} \cdot Av_{i,t} \le P_i^{\max}; \qquad \forall s, \forall i, \forall t \tag{5.10}$$

$$Av_{i,t} = \left\{0, \forall t, \forall i \big|_{Trip_{i,t} \neq 0}\right\} \tag{5.11}$$

$$SOC_{s,i,t} \ge SOC_{i,t}^{Des} \quad \left\{\forall i, \forall t \big|_{Flag_{i,t}^{dep}=1}\right\} \tag{5.12}$$

Constraint (5.13) indicates total regulation capacity as the sum of the regulation up and down capacities. Equations (5.14) and (5.15) calculates respective regulation capacities (reg-up and reg-down) at a time t, provided by i^{th} EV. $POP_{s,i,t}^{\max}$ is the maximum value of charging rate and computed as per (5.16). Constraint (5.17) is the congestion management or overloading constraint indicating that transformer delivery capacity must be greater than or equal to the amount of combined charging rate, total regulation down capacities, and baseload at time t.

$$rc_{s,i,t} = uc_{s,i,t} + dc_{s,i,t} \quad \forall s, i, t \tag{5.13}$$

$$uc_{s,i,t} = \left(POP_{s,i,t} - P_i^{\min}\right) \cdot Av_{i,t} \tag{5.14}$$

$$dc_{s,i,t} = \left(\min\left(P_i^{\max}, POP_{s,i,t}^{\max}\right) - POP_{s,i,t}\right) \cdot Av_{i,t} \tag{5.15}$$

$$POP_{s,i,t}^{\max} = \left(SOC_i^{Des} \text{-SOC}_{s,i,t}\right) \cdot \text{Av}_{i,t} \cdot \frac{\text{BC}_i}{Ef_i} \tag{5.16}$$

$$TDC \ge \sum_{i=1}^{N} \left(POP_{s,i,t} + dc_{s,i,t}\right) + l_t^b \quad \forall t, \forall s \tag{5.17}$$

5.5 Results and Discussions

The scenario generation algorithms have been realized for subsequent-day market prices of NYISO [134]. The outcomes verify the capability of the anticipated algorithms in developing uncertain prices. These modelled algorithms can efficaciously be exploited to produce price offers for optimal power markets trading of EVA. Figs. 5.2 and 5.3 indicate reduced scenarios of regulation prices and electricity prices along with the forecasted prices.

The resulting cumulative distribution functions for the profit in cases $\beta = \{0,1\}$ are represented in Figs. 5.4 and 5.5. For $\beta = 0$, the $CVaR = -1209.11$ (\$) and Expected profit = 116.1 (\$) whereas for $\beta = 1$, their values are equal to -99.76 (\$) and 15.43 (\$). Figs. 5.6 and 5.7 show the efficient frontiers as value-at-risk and conditional value-at-risk versus expected profit, obtained from variation of β from 0 to 1 respectively for risk-neutral and risk-averse EVA. Different schedules are obtained with reduced expected profit for risk-aversive EVA as compared to risk-neutral EVA. It depicts that *VaR* and *CVaR* increases for a higher value of the risk-weighting parameter. Baseload profile is coordinated from SO to EVA. Fig. 5.8 illustrates aggregated charging load variation in coordination with baseload profile. Due to TOU-PBDR, valley filling is observed during off-peak hours and charging load is lower during peak hours.

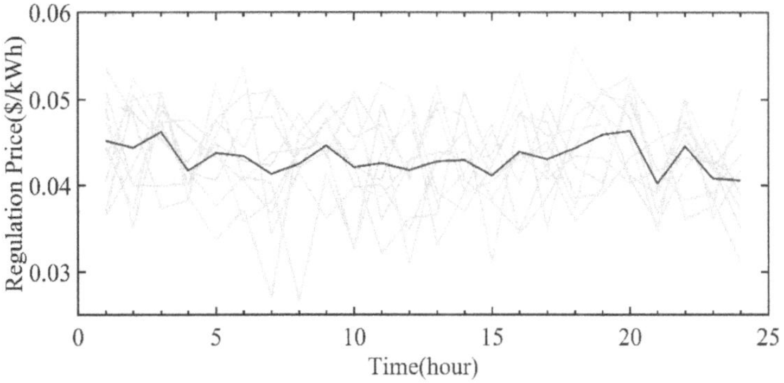

Fig. 5.2: Scenario Reduction of Regulation Price (Green) along with the Forecasted Regulation Price (Blue)

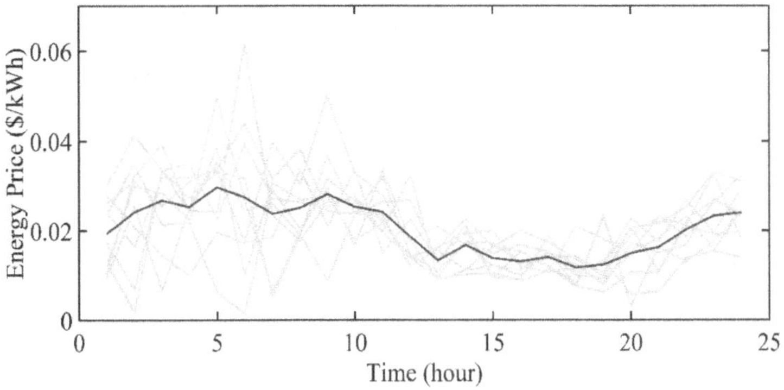

Fig. 5.3: Scenario Reduction of Energy Price (Green) along with Forecasted Energy Price (Blue)

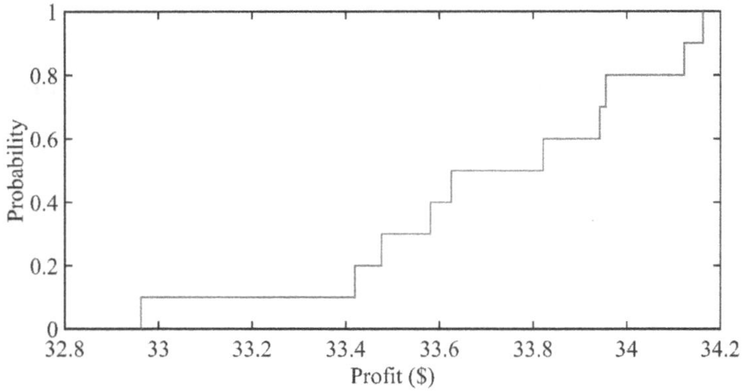

Fig. 5.4: Cumulative Probability Distribution Function for $\beta = 0$

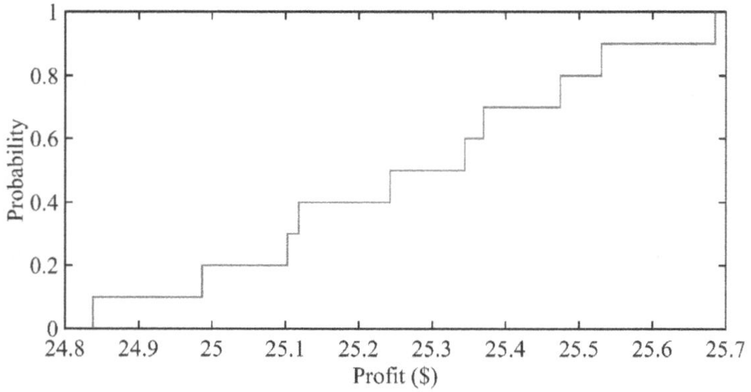

Fig. 5.5: Cumulative Probability Distribution Function for $\beta = 1$

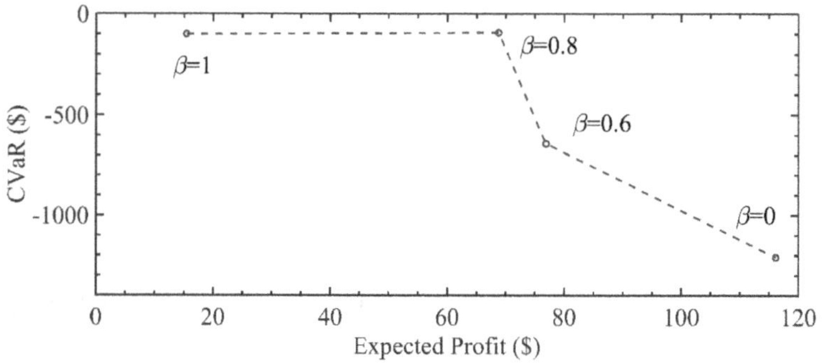

Fig. 5.6: Efficient Frontier of *CVaR* Versus Expected Profit

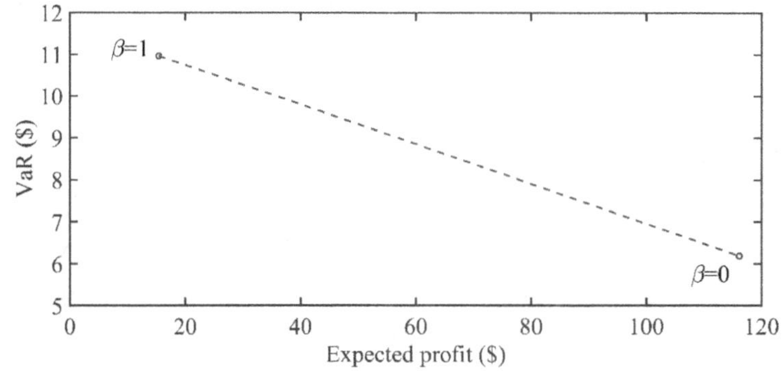

Fig. 5.7: Efficient Frontier of VaR Versus Expected Profit

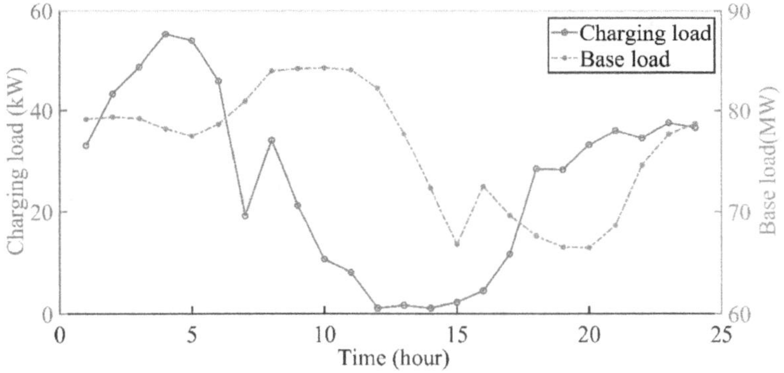

Fig. 5.8: Aggregated Charging Load Versus Baseload on the Grid

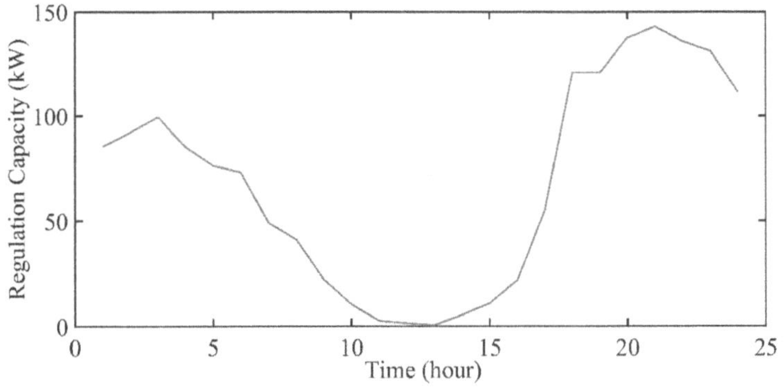

Fig. 5.9: Total Regulation Capacity at an hour t

When EVA becomes more risk-aversive, it modifies its scheduling to increase sell of regulation services to SO, as shown in Fig. 5.9 and energy selling in the market for off-peak duration. Its participation in the regulation market is more as compared to the energy market, which affects its profit positively. In deregulated milieu, the risk-aversiveness issue could have a substantial impact on EVA's decision-making for executing TOU-PBDR and smart coordination.

5.6 Epilogue

Proposed work models an optimal risk-constrained static G2V scheduling of EVA incorporating TOU-PBDR to maximize its profit while supplying energy requirement of EV owners. The chapter presents a financially viable optimization basis for aiding the smart charging of EVs. The risk assessment model for financial risk management considered energy and regulation prices uncertainties whereas *CVaR* for the risk estimation. Stochastic optimization problem formulation is utilized for comprehensive techno-economic evaluation through MCS based scenarios generation and KD based reduction. Results show that, for risk-averse EVA, different schedules obtained with more costs as compared to risk-neutral EVA. As a part of future work, risk-based scheduling of EVA for more pragmatic G2V could be incorporated to investigate the impact of multiple uncertainties viz. EVs' mobility behaviour uncertainty and market price uncertainty.

Chapter 6

RISK-AVERSE INTEGRATED DR AND DYNAMIC G2V SCHEDULING OF EVA FOR MULTIPLE MARKETS CONSIDERING GRID STABILITY

As discussed in Chapter 4, EVA trades in multiple markets through G2V charge scheduling. It provides regulation services to SO for grid stability through coordinated and distributed charging of EVs. However, against flat rate, PBDR integrated scheduling motivates EVs to charge in off-peak hours for load levelling, peak reduction and charging cost reduction as well. RTP based DR, mostly adopted in literature, has a low acceptance rate by EVs for being too dynamic to respond and being infeasible for few charge cycles in a day. TOU PBDR, being less dynamic, is a natural price signal for EVs and shields the EV owners from RTP volatility. In this regard, the TOU-PBDR model for EVA's charge scheduling is formulated for the energy and regulation market. TOU price is developed by EVA from RTP using AHC. Such scheduling is subjected to uncertainties like EVs' mobility behaviour and market prices, affecting severely its G2V operational behaviour. The work is extended with risk-averse charge scheduling modelling through a stochastic programming framework for multiple uncertainties. In the previous chapter, risk-based G2V scheduling of EVA is considered under an irrational static scenario in which EVA has complete charging profile information from EV owners $\left(EV\ ID,\ ta_i,\ td_i,\ SOC_i^{ini},\ SOC_i^{des}\right)$ prior ahead of the operating timeslot. However, in a more realistic, pragmatic and naturally wiser to implement DCSP, charging profile information of EVs is prior not stated to EVA and is uncertain. This chapter works on Integrated TOU-PBDR and Dynamic G2V Charge Scheduling of an EVA to support grid stability by selling regulation services to SO. The objective is to maximize the profit of EVA by the sale of up and down regulation power and charging of EVs and minimize the charging cost of EV owners. This multi-objective problem is subjected to the uncertainty of energy and regulation market prices and ta_i, td_i of an ith EV. In an optimization problem, the uncertain input variable calls for modelling of the problem considering it. Such problems can be modelled through a stochastic optimization approach where variable uncertainty is handled by various probable sample values called scenarios. Monte Carlo Simulation (MCS) approach is an effective stochastic optimization approach which gives precise result for a large number of scenarios. Nevertheless, analysing the aforementioned set leads to the intractable problem along with huge computing times. Considering this, an effective scenario reduction algorithm is necessitated which can retain the pattern of uncertainty. Kantorovich Distance (KD) matrix scenario reduction technique is one of the prevalent schemes and is employed in the present work. It deletes and aggregates the scenarios by KD matrix, which is the likelihood distance between two different scenario subsets that characterises the comparable stochastic procedure.

The input parameter uncertainties pose the risk of the optimality of objectives of the problem and necessitate financial risk management. $CVaR$ is employed widely in problems related to electricity markets due to its linear formulation and also being a coherent risk measure. Therefore, the research work is further extended to model the risk-averse integrated TOU-PBDR framework of EVA to manage its revenue risk and reduce energy purchase cost considering mobility behaviour dynamics. It could meet schedule plans and manage the adverse effect of market price and times of arrival and departure uncertainties. With CVaR as a risk measure, results illustrate a trade-off among profit and risk with optimal EV charging cost, EVA profit and regulation provision to SO, considering realistic case studies.

The rest of the chapter organization as follows. A brief introduction to the necessity of risk-based formulation in dynamic G2V scheduling is highlighted in Section 6.1 and related work is discussed to maintain the continuity of the work in Section 6.2. The theoretical basis elaborates features of the proposed model in Section 6.3. The sub-sections of this Section 6.3 describe respectively the market framework, uncertainty categorization through scenarios generation/reduction, and implementation steps of risk controlling for risk-management in stochastic optimization. The problem formulation of the risk-neutral and risk-averse EVA's dynamic G2V scheduling strategy incorporating TOU-PBDR is proposed in Section 6.4. The case studies with simulation outcomes are presented and investigated in Section 6.5. The concluding remarks are made in Section 6.6.

6.1 Introduction

EVA computes a charging schedule with a DCSP algorithm, and an individual EV owner obeys it for charging operation after its arrival. There can be two approaches to resolve the problem of charge scheduling. In the primary technique, the EVA solves the problem to determine the charging schedule of EV i every time an EV arrives at EVA for charging task. After computing this charging schedule, EVA does not modulate it during complete charging hours. It is faster due to the smaller size of the problem with reduced constraints and variables. However, it is not so accurate in performance. Therefore, to improve accuracy, another approach of dynamic G2V planning along with rectifying schedules in which EVA modulates charging rates of plugged-in EVs whenever m new EVs arrive. The aforementioned could be performed through consideration of EVs in particular at present plugged into the system. The smaller value attributed to m provides better performance in terms of a trade-off between computational burden and accuracy. As discussed in Chapter 4, Section 4.4, EVA specially designs TOU rates for EV charge scheduling for providing more flexibility to EV owners for charging cost minimisation and avoid the rebound effect through distributed charging over connection hours. As mentioned earlier in Chapter 5, Section 5.1, EVA faces market price risk in G2V scheduling by buying energy at RTP and selling at TOU price. Due to the inherent volatility of RTP, EVA may incur a loss. It necessitates financial risk management for EVA's DCSP framework integrating TOU-PBDR, with mobility behaviour uncertainty as well as multiple market price uncertainties associated with energy and regulation markets.

As stated in Chapter 5, $CVaR$ is employed widely as an efficient risk quantifier for downside risk, more conservative than VaR due to consideration of risks beyond VaR. It is a coherent risk quantifier and could be calculated efficiently considering all symmetric as well as non-symmetric distributions. It provides stable statistical estimates and shows consistency at different confidence levels compared to VaR. Similar to chapter 4 and 5, the MCS technique is imitated here to model the uncertain parameters of market prices pattern and EV owners' arrival and departure behaviour [118-119]. KD matrix scenario reduction technique is one of the prevalent schemes and is employed in the present work for EV mobility behaviour and market prices scenario reduction [119, 126]. It deletes and aggregates the scenarios by KD matrix, which is the likelihood distance between two different scenario subsets that characterises the comparable stochastic procedure. The flow diagram of the uncertain parameter's scenario generation as well as reduction, and the mathematical steps for scenario reduction can be referred from [121] or Chapter 4, Section 4.3.

In this context, the proposed work presents a stochastic optimisation model for risk-averse integrated TOU-PBDR and dynamic G2V charge scheduling of EVA in the day-ahead multiple markets viz. energy and regulation service provision. EVA participates in these multi-market aiming at profit maximisation through dynamic charge scheduling. The times of arrival and departure of EVs and hourly market prices are considered as potential uncertainties, which are represented in MCS scenarios. The conditional value at risk ($CVaR$) is utilised as a risk quantifier for managing EVA's financial risks caused by the uncertainties. Risk-averse EVA's dynamic G2V scheduling strategy is formulated as a DNLP in GAMS. CONOPT solver is used for solving the resulting DNLP problem. Accordingly, EVA maximises its expected profit and $CVaR$ in the risk-averse decision-making problem while minimising the charging cost from EV owners' perspective.

This chapter presents a risk-averse TOU-PBDR integrated dynamic G2V scheduling framework for an EVA:

1) to decide the TOU pricing for EV owners from forecasted RTP data for implementing TOU-PBDR.

2) to execute portfolio optimisation of various market products like energy, regulation up and regulation down capacities to hedge against market price risks and mobility behaviour dynamics.

3) simultaneously maximise the expected profits and $CVaR$ through a developed stochastic optimisation algorithm

4) In this trade-off, $CVaR$, the most preferred risk quantifier in fiscal risk assessment, is employed for risks modelling.

6.2 Theoretical Basis

6.2.1 Framework

The G2V charging network architecture adopted is the same as Chapter 3. The combined EVs are coupled to the leaf node transformers guaranteeing that the delivery capacity of each of the leaf node transformers is not exceeded by regulated EV charging rate to avoid the congestion of transmission lines and other equipment. An EVA may charge the EVs at any suitable rate of charging based on battery capacity and the load conditions, at off-peak load with higher charging rates and vice versa. Simultaneously accomplishing these responsibilities, EVA also endeavours to revenue optimization.

It is assumed that EVA has a call for the supply of regulation services from SO. In the first stage, EVA determines the TOU sale price of electricity for EVs from the forecasted RTP. As stated, TOU tariffs are formulated by EVA for season, weekdays and weekends and conveyed to the EVs beforehand. EVA charge scheduling is an operational decision-making problem. Henceforth, it determines the optimal charging rate of EVs as per their probable arrival and departure times and probable initial SOC to maximize its profit from the sale of regulation to SO, sale of energy to EVs at predetermined TOU and purchase of energy from the energy market at RTP. Based on different battery capacities and consumption patterns, there can be different groups of contracted EV customers. The scheduling horizon is usually split into several periods, according to the times for which decisions are made. Here, hourly scheduling is used.

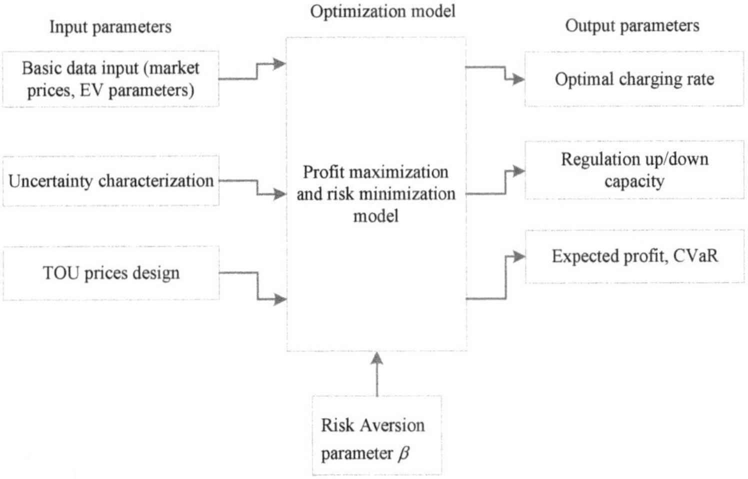

Fig. 6.1: Risk-Averse EVA's Dynamic G2V Charge Scheduling Framework Incorporating TOU-PBDR

Fig.6.1 illustrates the Integrated TOU-PBDR and dynamic G2V scheduling framework of risk-averse EVA. It is assumed that the EVA being a price-taker submits only quantity supply/demand in energy and regulation markets. Thus, the market prices are independent of the EVA's actions. Optimal charge scheduling of an EVA gives the scheduled charging rate of an EV which is also its POP. $POP_{i,t}$ is an average level of charging power drawn by an ith EV over t^{th} an hour while providing regulation capacity. The regulation capacity of an EV is the amount of charging rate that it can increase or decrease around $POP_{i,t}$ as discussed in Fig. 3.2.

EVA procures electricity at RTP through wholesale energy market participation. In this work, charge scheduling of EVA is developed for weekdays of the summer season. Hence, TOU for such days is designed based on forecasted RTP. The data for RTPs of all the weekdays of a typical summer season month (September 2018), from the New York Independent System Operator (NYISO) [134] is taken and a forecasted RTP is generated using normal distribution curve fitting. This forecasted RTP is shown in Fig. 6.2. The hour axis is graded from 1 to 24 hours (i.e. 1 am to 12 am). It can be observed that RTP is higher during 12-23 hours and lesser during the 24th hour and 1-11 hours. In this work, EVA devises a five-tiered TOU for EV owners from forecasted RTP. Five tiered TOU is selected to provide more flexibility to EV owners. Noticeably a well-accepted TOU design method of AHC has been adopted [78].

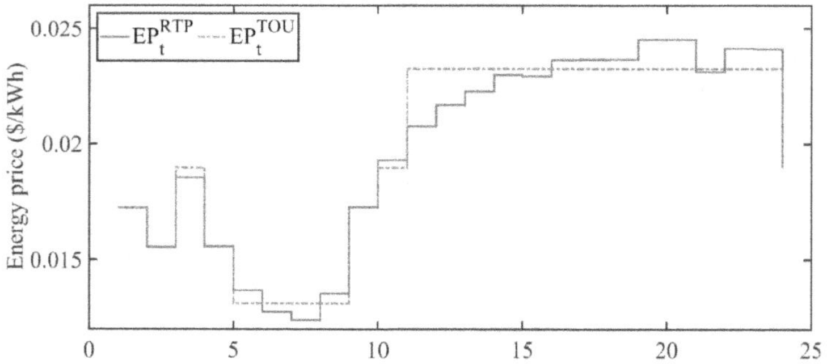

Fig. 6.2 RTP of energy in Capitol Zone of NYISO and designed TOU rates.

For a small-sized dataset, AHC is preferred due to inherent accuracy. It is an effective algorithm generating wide clusters by comprehending the price pattern. Based on the number of clusters, the average within-group distance (Euclidean distance) is computed. Lesser average within-group distance specifies higher similarity inside the group. Five clusters are determined for five-tier TOU i.e. off-peak, shoulder off-peak, flat, shoulder peak, and peak. AHC gives the length of each cluster as such the typical hours of RTP falling in them. Hence, the TOU tariff of a cluster is the average of the RTP of the hours in it.

$$\text{TOU rate for cluster } k = \frac{\text{Sum of forecasted RTPs in the cluster}}{\text{Number of hours in the cluster}} \quad (6.1)$$

EVA would add a mark-up value to designed TOU rates and retail electricity to EV owners to charge EVs. The focus of the paper is to execute TOU-PBDR and dynamic charge scheduling of EVA, considering multiple uncertainties. The risk of profit variation is modelled and minimized along with profit maximization of EVA. Through TOU-PBDR, the EV owner is motivated to deviate high charging during off-peak hours.

6.2.2 Uncertainty Characterization

An EVA faces multiple uncertainties where energy and ancillary service market prices and mobility behaviour of EVs are the most significant ones. Possible capturing of joint realisation of these uncertainties at every timeslot over the scheduling horizon is necessary. Here, it is supposed to have an EVA with 1000 EVs and the battery capacity of each EV is 80 kWh. The EVs characteristics are obtained from [121], and the range of their arrival and departure times are extracted from [93]. For an ith EV, driving characteristics are randomised ta_i and td_i, and SOC_i^{ini} during arrival. For uncertainty modelling, EV's ta_i and td_i are assumed to follow normal Probability Distribution Function (pdf) with forecasted value as mean (μ_{ta_i} and μ_{td_i}) and a standard deviation (σ_{ta_i} and σ_{td_i}). EVs are connected to the grid for charging during late evening and night hours due to travel and office hours during the day. Thus, over 1-24-hour horizon (μ_{ta_i} and μ_{td_i}) are 19:00 hour and 7:00 hour respectively and (σ_{ta_i} and σ_{td_i}) are taken as 2 hours.

Table 6.1 Designed Five-tiered TOU rates from Forecasted RTP

	Off-Peak	Shoulder Off-Peak	Flat	Shoulder On-Peak	On-Peak
Duration	t_5-t_8	t_2, t_4	t_1, t_9	t_3, t_{10}, t_{24}	t_{11}-t_{22}
TOU rates ($/kWh)	0.013	0.0156	0.017	0.019	0.023

EV's SOC_i^{ini} specifies the aggregated impacts of battery usage based on travelled distance and any other trip losses. Therefore, SOC_i^{ini} scenarios lie in fine range (10% to 30% of battery capacity) reflecting the worst-case scenarios, considering all possible driving pattern losses. SOC_i^{des} at departure is fixed to 90%. Uncertainty of energy and regulation market prices is also modelled as normal pdf. The mean and standard deviation of the hourly historical prices over weekdays of the summer season (September 2018) from NYISO [134] is used as parameters of pdf.

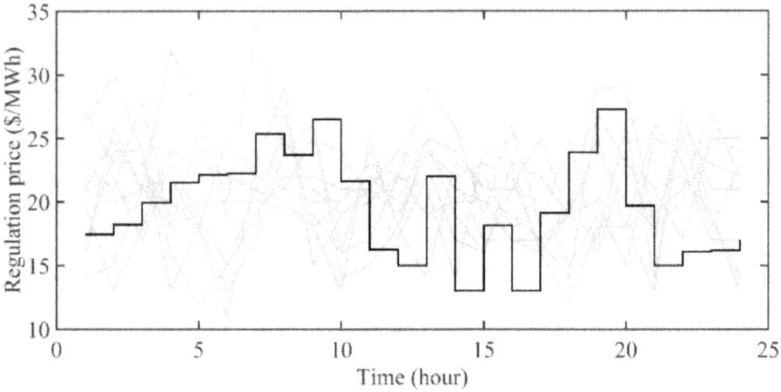

Fig. 6.3 Reduced Scenarios of the regulation price(green) with the forecasted regulation price (black)

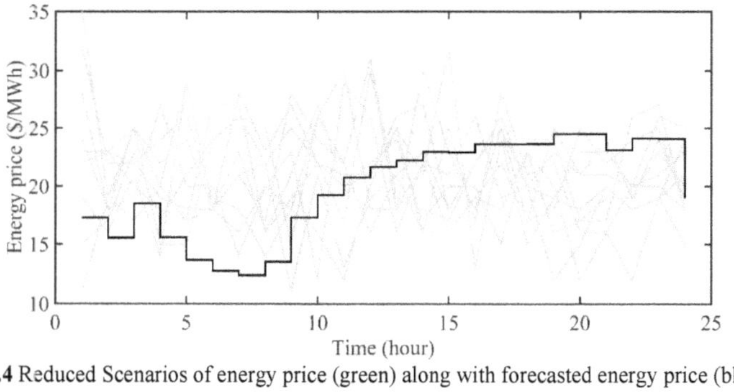

Fig. 6.4 Reduced Scenarios of energy price (green) along with forecasted energy price (black)

As mentioned in Section 6.1, this chapter employs MCS based stochastic optimization approach for a solution. In this approach, the optimization of the expected value of the objective function over scenarios of uncertain parameters is conducted. A large number of scenarios of uncertain parameters is generated using normal pdf and KD matrix-based scenario reduction technique is used for the reduction of generated ones into a small number which can result in a lesser computational burden for the problem. KD method has two types: backward reduction and forward selection. The forward selection method requires higher CPU times while backward reduction algorithms are faster and efficient. KD based backward method is used in this chapter. A thousand scenarios of RTP of the energy market and regulation market and mobility behaviour of each of the ith EV are generated which are then, reduced to ten scenarios. Fig. 6.3 and Fig.6.4 shows the same respectively for the regulation market and RTP of energy.

To evaluate the expected objective function, N_s scenarios of ith EV are created where a scenario s , $s \in N_s$, is represented by equation (6.2).

$$Scenario\ s = \left\{ ta_{i,s}, td_{i,s}, EP_{s,t}^{RTP}, RP_{s,t} \right\} \ \forall t \in T, i \in N \tag{6.2}$$

The probability of occurrence of such scenario s of ith EV, $prob_{i,s}$, is the product of the respective probabilities of each uncertain parameter scenario probability, as given by equation(6.3)

$$prob_{i,s} = \prod_{i \in N, t \in T} \left(\pi_{s-ta_i} \times \pi_{s-td_i} \times \pi_{s-EP_t^{RTP}} \times \pi_{s-RP_t} \right) \tag{6.3}$$

Here π_{s-ta_i}, π_{s-td_i}, $\pi_{s-EP_t^{RTP}}$, and π_{s-RP_t} are the probability of individual scenarios of ta_i, td_i, EP_t^{RTP} and RP_t respectively.

EVA aims at maximisation of its expected profit under uncertainties through satisfying charging requirements of contracted EV owners and regulation services to SO. To this point, it participates in the wholesale energy market as a buyer and further as a seller in the regulation market with a capacity of connected EVs. The profit in the current framework is thus determined as the revenue from the sale of energy to EV owners and regulation up/down services to SO minus the purchase costs of charging requirements from EVs.

6.2.3 Risk Management

Risk is an important concept in the G2V decision-making problem of EVA and can be defined as a degree of uncertainty in profit attained. While scheduling the G2V charging of EVs, the decision of EVA should consider the minimization of the risk of profit variation along with the profit maximization. As discussed in Chapter 5, for financial risk management of EVA in uncertain conditions, this chapter utilises $CVaR$ a risk measure for it being popular due to its linear formulation and coherency. Fig. 6.5 shows VaR and $CVaR$ by a probability distribution of profit under α confidence (risk) level. For a given α between (0, 1), $CVaR$, is the expected value of the profit smaller than the $(1-\alpha)$ quantile of the profit distribution. Mathematically, the $CVaR$ for a discrete distribution is defined as equation (6.4):

$$CVaR(\alpha,\mathrm{X}) = \max\left\{\eta - \frac{1}{1-\alpha} E_s\left\{\max\left\{\eta - f(X,s),0\right\}\right\}\right\}, \forall \alpha \in (0,1)$$

$$\varphi(s) \geq (\eta - f(X,s)), \varphi(s) \geq 0; s = 1,2,\cdots,N_S$$

(6.4)

Here, $f(X,s)$ is a profit function in the scenario s and $\varphi(s)$ is an auxiliary variable [36]. Maximization of $CVaR$ is risk minimization. As profit and risk are conflicting objectives, a multi-objective formulation $Max\ ((1\text{-}\beta)\Pr ofit + \beta \cdot CVaR)$ can be adopted where $\beta \in [0,1]$. β represents the risk aversion parameter or a weighing parameter employed to realise the trade-off between expected profit and risk aversion. Its value is 0 and 1 denotes respectively risk-neutral and risk-averse decisions. Optimal η is VaR of the profit at a confidence level α. It is the highest value of profit so that the probability of the actual profit being lower than or equivalent to η is lower than or equivalent to $(1-\alpha)$. $CVaR$ represents the mean of scenarios' profits with earnings lower than VaR.

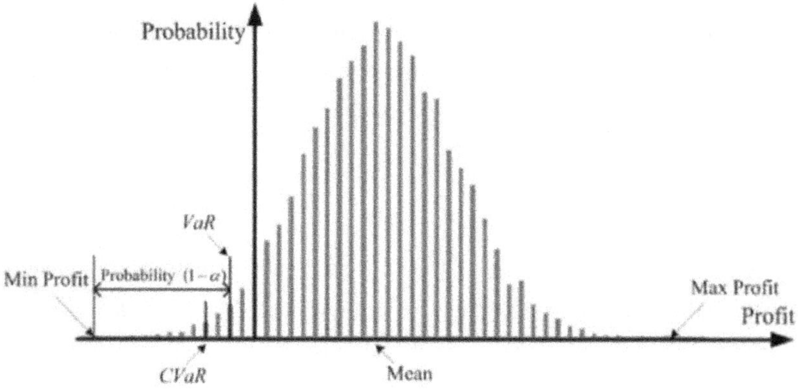

Fig. 6.5. Illustration of VaR and $CVaR$ [36].

6.3 Problem Formulation

The objective of the optimal Risk Aversion Decision Making (RADM) scheduling of EVA in multiple markets, with $CVaR$ a risk measure, can be formulated as a stochastic programming problem, given by (6.5)

$$\underset{POP_{i,s,t},uc_{i,s,t},dc_{i,s,t}}{Max(RADM)} = (1-\beta)\left[\sum_{i=1}^{N}\sum_{s=1}^{N_s} prob_{i,s}\cdot \Pr ofit_{i,s}\right] + \beta(CVaR) \tag{6.5}$$

$$CVaR = \eta - \left(\frac{1}{1-\alpha}\right)\sum_{i=1}^{N}\sum_{s=1}^{N_s} prob_{i,s}\cdot \varphi_{i,s} \tag{6.6}$$

The first term in RADM is the profit and the second one is $CVaR$, both are being trade-off through β. $\varphi_{i,s}$, β, and α are the auxiliary variable, risk aversion parameter, and confidence level respectively.

$$\varphi_{i,s} \geq \eta - \Pr ofit_{i,s} \tag{6.7}$$

$$\varphi_{i,s} \geq 0 \tag{6.8}$$

$$\Pr ofit_{i,s} = \operatorname{Re} venue_{i,s} - Cost_{i,s} \tag{6.9}$$

$$\operatorname{Re} venue_{i,s} = \sum_{t=1}^{T} RP_{s,t} rc_{i,s,t} + \sum_{t=1}^{T}\left(\begin{matrix}\left(M_{k_t} + EP_t^{TOU}\right)\\ POP_{i,s,t} h_{i,s,t}\end{matrix}\right) \tag{6.10}$$

$$\operatorname{Cos} t_{i,s} = \sum_{t=1}^{T} EP_{s,t}^{RTP}\cdot POP_{i,s,t}\cdot h_{i,s,t} \tag{6.11}$$

$$\Pr ofit_{i,s} = \sum_{t=1}^{T}\left[\begin{matrix}RP_{s,t}\cdot rc_{i,s,t} + \left(M_{k_t} + EP_t^{TOU}\right)\cdot \\ POP_{i,s,t} h_{i,s,t} - EP_{s,t}^{RTP} POP_{i,s,t} h_{i,s,t}\end{matrix}\right] \tag{6.12}$$

$$h_{i,s,t} = \begin{cases}1, & ta_{i,s} \leq t \leq td_{i,s}\\ 0, & otherwise\end{cases} \tag{6.13}$$

$$POP_{i,s,t}^{\max} = \left(SOC_i^{des} - SOC_{i,s,t}\right)\cdot h_{i,s,t}\cdot \frac{BC_i}{Ef_i} \tag{6.14}$$

The $h_{i,s,t}$ parameter is decided from values of $ta_{i,s}$ and $td_{i,s}$. It depicts the EV's plugged-in status. $POP_{i,s,t}^{\max}$ is the charging power required by i^{th} EV for reaching up to desired departure SOC from current SOC in one timeslot, as represented by (6.14). In constraint (6.15), a value is assigned to the SOC of each plugged-in EV at every timeslot. Scheduled charging rate $POP_{i,s,t}$ should lie within upper and lower limits as in (6.16). The upper limit of EV owners' charging cost (CB) could be determined from an unregulated baseline charge scheduling case [65]. The charging cost of an ith EV must be less than this CB as expressed by (6.17). Constraint (6.18) is the congestion management constraint preventing overloads of distribution transformer due to connected baseload, aggregated charging load and regulation down capacities at each timeslot. As per (6.19), the SOC of ith EV in any hour t must be less than or equal to its desired SOC. (6.20) and (6.21) respectively estimates the regulation up and regulation down capacities provided by an ith EV in an hour t. (6.22) calculates the total regulation capacity of ith EV in an hour t as the sum of up and down capacities.

$$SOC_{i,s,t} = \begin{cases} \left(SOC_{i,s,t-1} + Ef_i \cdot \frac{POP_{i,s,t-1}}{BC_i} \right) \cdot h_{i,s,t-1}; & ta_{i,s} < t < td_{i,s} \\ SOC_{i,s}^{ini}; & t = ta_{i,s} \\ SOC_i^{des}; & t = td_{i,s} \end{cases} \tag{6.15}$$

$$P_i^{\min} \cdot h_{i,s,t} \leq POP_{i,s,t} \leq P_i^{\max} \cdot h_{i,s,t} \tag{6.16}$$

$$\sum_{t=1}^{T} \left(M_t + EP_t^{TOU} \right) \left(POP_{i,s,t} \cdot h_{i,s,t} \right) \leq CB \tag{6.17}$$

$$\sum_{i=1}^{N} \sum_{s=1}^{N_s} \left(POP_{i,s,t} + dc_{i,s,t} \right) + l_t^b \leq TDC \tag{6.18}$$

$$SOC_{i,s,t} \leq SOC_i^{des} \tag{6.19}$$

$$dc_{i,s,t} = \left(\min\left(POP_{i,s,t}^{\max}, P_i^{\max} \right) - POP_{i,s,t} \right) \cdot h_{i,s,t} \tag{6.20}$$

$$uc_{i,s,t} = \left(POP_{i,s,t} - P_i^{\min} \right) \cdot h_{i,s,t} \tag{6.21}$$

$$rc_{i,s,t} = uc_{i,s,t} + dc_{i,s,t} \tag{6.22}$$

The proposed problem formulation, RADM, is the expected profit maximisation of EVA under stochastic parameters. In this work, the EVs owners' optimal charge scheduling based on TOU price rates offered by EVA is determined where EVA is subjected to the volatility of RTP and regulation price and mobility behaviour of EVs. The risk of profit variation of EVA is also handled. The risk aversive nature of EVA varies from $\beta = 0$ (Risk Neutral) to $\beta = 1$ (Risk-Averse).

6.4 Simulations and Results

A sample case study of 1000 EV owners coming to an EVA for charging is adopted. The charging efficiency and battery capacity of individual EV are respectively 90% and 80 kWh. The lower and upper limits of the charging rate are considered as 0 kW and 8.8 kW respectively, applied in the simulation of the G2V algorithms. The travelling behaviour of EVs has been taken from the National Household Travel Survey (NHTS) database [93]. As given in section 6.2.1 and 6.2.2, five tiers TOU is formulated by the EVA from forecasted RTP. Table 6.1 and Fig. 6.2 are to be referred for this case study. Also, the scenarios of RTP and regulation price for uncertainty modelling are given in Fig. 6.3(a) and Fig. 6.3(b). The proposed EVA scheduling is formulated as a Nonlinear programming problem in the General Algebraic Modelling System (GAMS) v24.2.3 and solved using CONOPT DNLP solver [137]. Three cases are considered as summarised in Table 6.2. Case 1 is the RNDM problem with RTP and regulation market price uncertainty modelling. Case 2 is a RADM problem considering risk management with Case1. Case 3 is case 2 with EVs' mobility behaviour uncertainty modelling. The value of β in case 2 and 3 is varied from 0 to 1 and the confidence level α is kept as 0.9.

Tables 6.3 and 6.4 show the results for the cases mentioned in Table 6.2. It is clear from tables that risk minimization consideration offers lesser expected profit than risk-neutral case. It also reduces the charging cost for EV owners. Price fluctuations in the energy and regulation market increase the total risk to EVA. In such circumstances, risk constrained scheduling of EVA suggests scheduling less in these markets. Thus, optimal scheduling reduces $POP_{i,s,t}$ to adapt to handle the risk of profit variation.

Table 6.2 Summary of Numerical Cases

Numerical cases	RTP and regulation market price uncertainty	EVs' Mobility behaviour uncertainty	Risk Management
Case1	√	x	x
Case2	√	x	√
Case3	√	√	√

Table 6.3 Expected Cost, Revenue and Profit for Case1and Case2

	RNDM $\beta = 0$ **(Case 1)**	**RADM** $\beta = 1$ **(Case 2)**
EV Owners' Expected Charging Cost ($)	1110.12	911.1167
EVA's Expected Revenue ($)	2548.936	2400.782
EVA's Expected Profit ($)	1674.514	1604.688

Table 6.4 Expected Cost, Revenue, and Profit for Case 3

	RNDM $\beta = 0$	RADM $\beta = 1$
EV Owners' Expected Charging Cost (\$)	1087.92	1090.801
EVA's Expected Revenue (\$)	2624.304	2608.311
EVA's Expected Profit (\$)	1846.831	1439.77

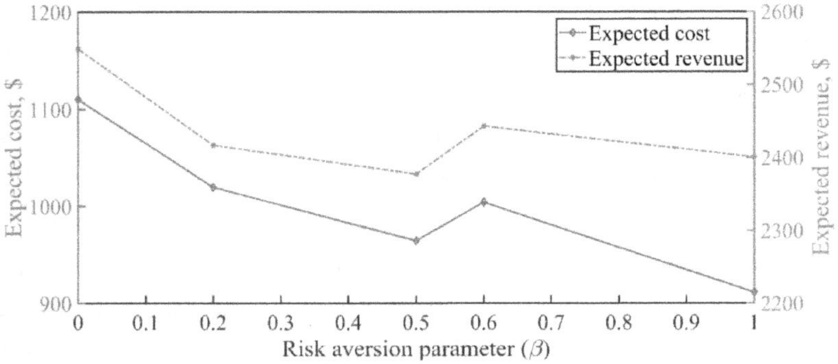

Fig. 6.6 Expected revenue and cost in case2 for EVA with varying β

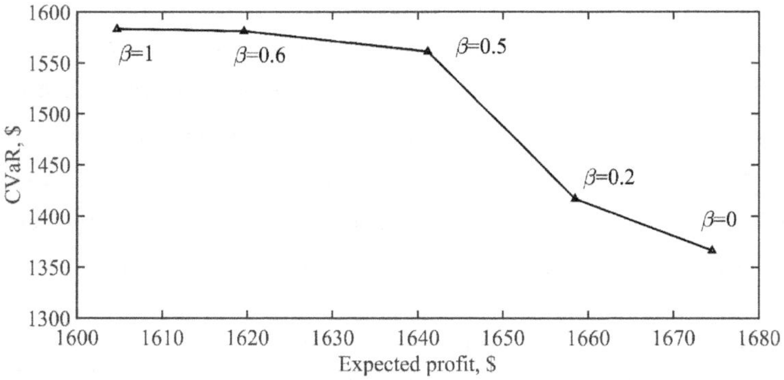

Fig. 6.7. Illustrates the efficient frontier for case 2

Fig. 6.6 shows the variation of expected energy procurement cost and expected revenue of EVA with risk aversion parameter β. As β increases, the charging rate $POP_{i,s,t}$ value decreases lowering the expected energy procurement cost of EVA, charging cost of EV owners and expected composite revenue (composed of charging revenue and regulation revenue).

The variation of risk measure $CVaR$ and expected profit for case 2 is shown in Fig. 6.7. This curve is a Pareto front between the two conflicting objectives. As the high value of $CVaR$ represents high-risk averseness, it is clear from the graph that as β rises, $CVaR$ rises and expected profit decreases. This graph shows that when the risk control is not considered $(\beta = 0)$, EVA faces a higher risk of profit variation but has a high expected profit in return. With a rise in risk aversion parameter β, expected profit becomes less significant as compared to RNDM case resulting in reduced expected profit. At $\beta = 1$, EVA earns a minimum expected profit with minimum risk of expected profit variation. EVA decides on the value of β depending upon its appetite of risk handling. A risk-averse EVA selects a large value for β and vice versa. EVA utilises the Pareto front of risk and expected profit as an instrument to resolve the trade-off.

Fig. 6.8, Fig. 6.9 and Fig. 6.10 depicts the hourly variation of aggregated charging load or charging rate in coordination with β, baseload profile and TOU prices respectively. It can be seen from Fig. 6.8, that as EVA becomes risk-averse from risk-neutral behaviour, it reduces charging load. Fig. 6.9 and 6.10 clearly show the advantage of TOU-PBDR in terms of charging of EVs during off-peak hours with high charging rates as compared to peak hours. Such charging helps in valley filling of system baseload. In this work distributed charge scheduling is adopted to maintain the grid stability and avoid the overloading of the transformer/grid by fast or simultaneous charging. As mentioned in section 6.2, the (μ_{ta_i} and μ_{td_i}) are 19:00 hour and 7:00 hour respectively and (σ_{ta_i} and σ_{td_i}) are 2 hours. Due to mean arrival time lying during peak hours and due to the distributed hourly charging algorithm, the EVs coming with low SOC (0.3) tend to charge during the remaining peak hours (in this case $\left(t_{19} - t_{24}\right)$) with a significant charging rate. The distributed charging leads to the rise of SOC of an EV after each hour and hence, its charging rate reduces near the departure time at which the EV attains 0.9 SOC. To maintain the distributed charging, an EV is constrained to attain 0.9 SOC in the departure hour only. It is finally observed in this work that distributed charging may lead to the charging of an EV during peak hours contradict the TOU-PBDR. However, TOU-PBDR shields the EV owners from the RTP volatility and reduces its overall charging cost.

From Figs. 6.11 and 6.12, it can be seen that a risk-neutral EVA offers higher regulation up capacity and lower regulation down capacity as compared to risk-averse EVA. Hourly charging rates are higher for an ith EV of risk-neutral EVA as compared to that of risk-averse EVA. On this basis, Fig. 6.2 and equation (6.14) and (6.20) clarify that a high hourly charging rate of an ith EV leads to low $POP_{i,s,t}^{\max}$ and hence low regulation down capacity for risk-neutral EVA. Similarly, equation (6.21) clarifies that for high charging rates, regulation up capacity would be high.

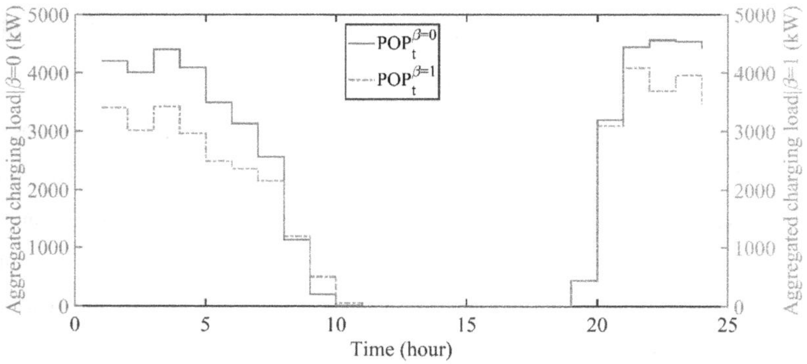

Fig. 6.8 Aggregated charging load for case2(β =0 and1)

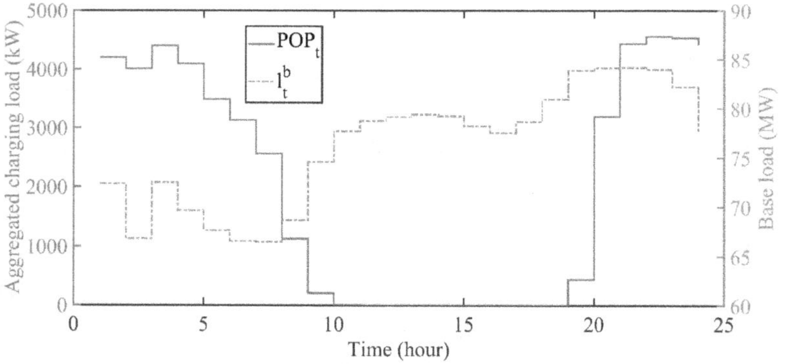

Fig. 6.9 Aggregated charging load versus baseload for case2(β =0)

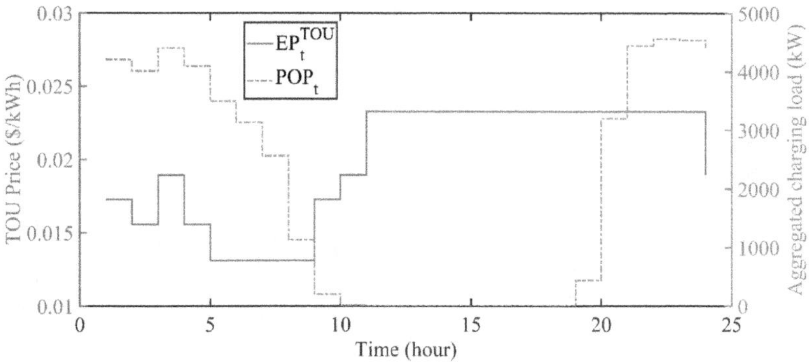

Fig. 6.10 TOU price and aggregated charging load for case2(β =0)

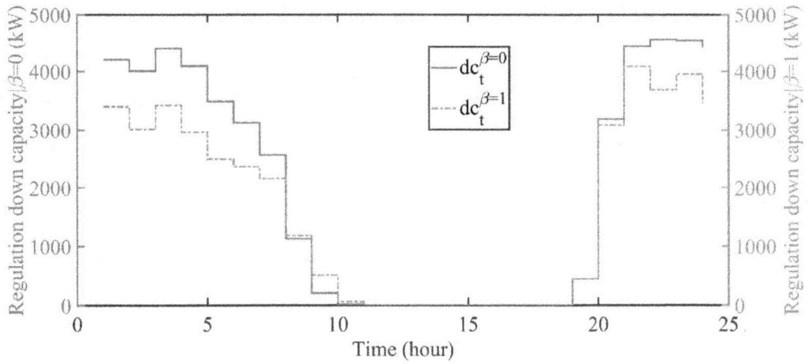

Fig. 6.11 Regulation down capacity for distinct timeslots with varying β for case2

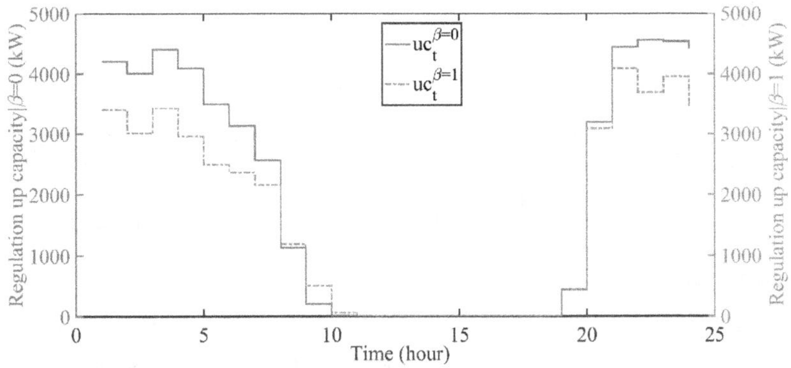

Fig. 6.12 Regulation up capacity for distinct timeslots with varying β for case2

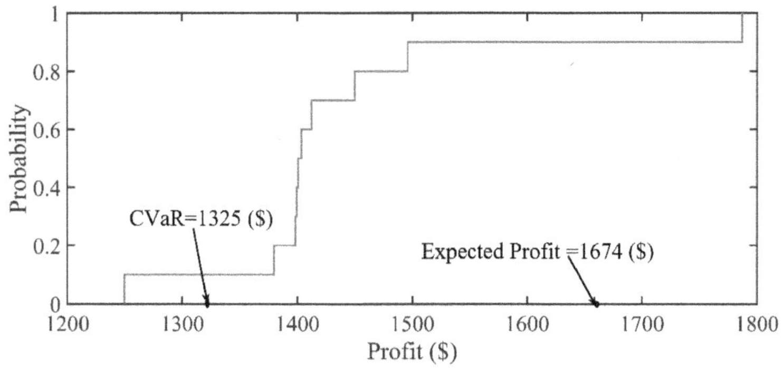

Fig. 6.13 Cumulative Probability Distribution of Profit for $\beta = 0$ in Case2

Figs. 6.13 and 6.14 represent the cumulative distribution function (c. d. f) of the profit for $\beta = \{0,1\}$. Since the risk-neutral problem only focuses on maximizing the expected profit, the risk of experiencing low profits in some scenarios is not avoided. For $\beta = 0$, the expected profit and $CVaR$ are respectively equal to \$1674 and \$ 1325. For $\beta = 1$, these are equal to \$1604.7 and \$1583.3 respectively.

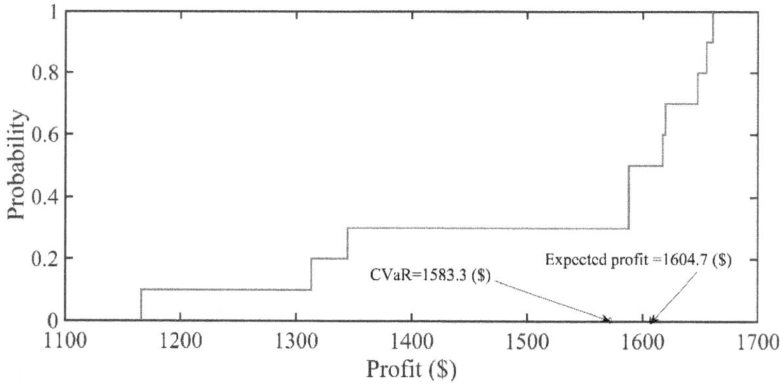

Fig. 6.14: Cumulative Probability Distribution of Profit for $\beta = 1$ in Case2

6.5 Epilogue

The dynamic mobility behaviour of EVs proposes extensive threats in the operations of the smart distribution system and power market. Proposed work models risk-averse dynamic G2V scheduling of EVA incorporating TOU-PBDR to maximise its profit while supplying energy requirement of EV owners. The risk assessment model for financial risk management considered uncertain parameters as EVs mobility behaviour and energy and regulation prices. *CVaR* is incorporated as a risk quantifier due to its inherent coherency for the risk control of unfavourable scenarios. MCS based stochastic optimisation problem formulation is utilized for comprehensive techno-economic evaluation. Results show that for risk-averse EVA, charge schedules are obtained with less charging rates as compared to risk-neutral EVA. Besides, risk-averse EVA provides more regulation down capacity and low regulation up capacity to SO than a risk-neutral EVA. The work can be extended with the inclusion of V2G in the charging algorithm of EVs of such a smart distribution network.

Chapter 7

INTEGRATED DR AND RISK-BASED DYNAMIC V2G SCHEDULING OF EVA FOR ENERGY AND ANCILLARY SERVICE MARKETS CONSIDERING GRID STABILITY

Electric Vehicles (EVs) are likely to be distributed energy storages and flexible price-based demand response (PBDR) resources for providing ancillary services to SO through Vehicle-to-Grid (V2G) scheduling. However, execution and transaction of integrated PBDR and V2G raise economic threats to V2G coordinator/an EVA in case of utility designed TOU prices. To resolve this, the proposed EVA specially designs TOU price from EV owners' perspective of charging cost minimization. TOU PBDR is integrated for improved market operations, to reduce charging costs of the EV owners and load levelling via enhanced flexibility to SO. As an interface between SO and EV owners, EVA felicitates smart coordination over operational V2G schedule in multiple markets for reliable regulation service provision. The multi-market operations of EVA providing energy and regulation services through bidirectional V2G scheduling are particularly challenging due to market price uncertainties. In this context, this chapter puts forward a risk-averse decision-making algorithm to maximize the expected profits of EVA offering regulation services to the SO for supporting grid stability and thus avoiding inevitable contingencies. The proposed algorithm considers multiple uncertainties of the energy prices, regulation market prices, and mobility behaviour using a stochastic programming framework. The pragmatic battery degradation modelling in the problem is capable of reducing battery ageing costs or increasing the lifetime of the EV batteries due to the smooth variation of charging-discharging rates, not accelerating the financial risk of battery ageing. It develops a pragmatic V2G scheduling algorithm integrating TOU-PBDR, accommodating EV battery degradation cost, adaptation of charging/discharging rate for each EV according to both load and prices. Economies of V2G operations, its resiliency and scalability to EVs' random mobility behaviour are validated through sensitivity analysis of EVA's performance parameters concerning battery replacement cost, round trip efficiency, number of EVs and upper limit of charging rate. The simulation results highlight the efficacy and viability of the proposed business model regarding peak load management and monetary benefits for all stakeholders viz SO, EV owners, and EVA.

The chapter is organized as follows. Section 7.1 highlights the necessity of DR integration in dynamic V2G scheduling and related work in this field. Section 7.2 discusses the system model for risk-averse V2G scheduling of EVA employing hierarchical control framework of EVA for the operational management of V2G involving operational management algorithm of EVA for grid integrated EVs, battery degradation model, TOU design for PBDR, and proposed optimal dispatch algorithm of EVA based on TOU prices and SOC level of EVs. Section 7.3 represents the charging/discharging rate modulation algorithm for EVA. Section 7.4 details the DR integrated V2G scheduling problem of EVA. Simulation results of DR integrated V2G scheduling are discussed in Section 7.5. The concluding interpretations are delivered in Section 7.6.

7.1 Introduction

Large-scale adoption of EVs would increase average electricity consumption and peak demand of the electrical grid. Also, uncoordinated charging of EV would make the situation more severe and require network reinforcement. One of the potential solutions to this problem is peak load management through EVs coordinated charging. In this regard, this chapter put forward the TOU price based V2G scheduling strategy of an EVA intending to optimize EVs coordinated charging/discharging and earn monetary benefits through multiple market participation viz. energy and regulation for all stakeholders involved such as EVA, SO, and EV owners. For this, TOU-PBDRP is proposed to enhance the value of EV owners and is executed in concurrence with SO mandates for improved load profile. Transformer overloading constraint, State-of-Charge (SOC) limits, and deviation of charging/discharging rate is considered for the pragmatic formulation of the V2G scheduling problem of EVA. The effectiveness of the proposed model is validated through peak load management and monetary benefits for all stakeholders. Moreover, this model offers regulation services to SO for supporting grid stability and thus avoid inevitable contingencies.

This chapter model the operational behaviour of EVA and its participation in TOU-PBDRP for providing energy and regulation services. To this end, a mixed-integer non-linear problem (MINLP) is formulated and the impact of TOU-PBDRP on the operational behaviour of EVA has been investigated. Market participation of EVA is optimized through appropriately designed TOU prices under a techno-economic environment. This work provides a useful template that EVA can adopt to maximize its benefits along with SO and EV owners through TOU Price based V2G scheduling, by capturing its baseline EV load, understanding the overloading risks, exploring load shaping solution with improved market operations, and thus, support widespread EV adoption while benefiting the grid.

7.2 Hierarchical Control Framework of Risk-Averse EVA for Operational Management of V2G

EVA provides a platform to EVs for active participation in electricity markets. It receives information from EV owners and coordinates with SO for load management. EV owners are instructed based on signals from SO, regarding modulation of charging/discharging rates and cycles. Local-level problems are handled by corrective actions (connection/disconnection) to provide flexibility to EV owners. Fig. 7.1 shows flexible hierarchical control architecture for V2G scheduling of EVA.

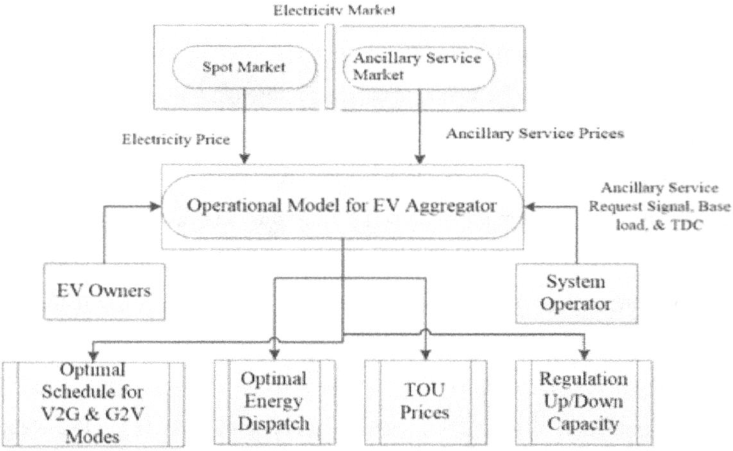

Fig. 7.1 A Schematic Block Diagram of Hierarchical Control Framework of EVA Illustrating V2G Scheduling

The proposed work deals with the problem of risk-based V2G scheduling of EVA in multiple markets considering the uncertainties of market prices and mobility behaviour in the ancillary service markets. Uncertainty characterization is done through a stochastic programming framework. The incorporation of TOU-PBDR motivates EV owners to charge their EVs in off-peak intervals in a distributed manner. Participation in energy and regulation markets increases the profit of EVA. SO's congestion management constraint and operational constraints of EVs are modelled in the proposed framework. *CVaR* is used as a coherent risk measure for controlling the risk of undesired scenarios. A realistic case study is considered to confirm the performance of the proposed model.

EVA as a dispatch controller can realize the economic value of V2G by controlling power flow between EVs and the grid [51]. Dynamic dispatch is performed by modification of dispatch according to the needs of the power system, to provide peaking power. EVs operate in a G2V mode according to their SOC requirements. A new revenue stream is generated through the provision of ancillary services like V2G regulation services to SO and incentivizing EV owners by reduced charging cost. Load levelling function by discharging/peak shaving on high demand periods and charging/valley filling on low demand periods. Thus, EVA stabilizes and smoothen fluctuations in the system by exhaustive utilization of V2G benefits. During power outages, EVA can support continuous power to run critical appliances through EVs. Electricity prices are low during off-peak hours. A higher value of POP is selected during such times to fill the valley. Therefore, EVA invigorates EV owners to charge during these opportune times. Subsidized rates to EV owners enable PBDRP to charge EVs fully before peak-demand hours. Minimum SOC is required for travel after work to ensure a return trip.

SO is benefitted by avoiding the use of costly generation during peak times while EV owners are benefitted due to compensation amount from EVA for the degradation of the battery. V2G management causes obscurity related to the life and capacity of the battery. To make V2G a viable and cost-effective solution, an efficient financial business model should be designed [160]. Economic analysis is necessary to determine whether V2G is a sensible solution or not. The willingness of EV owners to participate in V2G operation in response to TOU prices is the aim of EVA. Warranties regarding some battery lifecycles and operational constraints like SOC, DOD, etc. may attract EV owners. Thus, benefits must be balanced against risks faced by EV owners regarding battery lifecycle, EV availability, and ease of use. Frequent charge/discharge cycles impact battery lifecycle while rapid charge/discharge rates raise the temperature of the battery.

Fig. 7.2 illustrates the total regulation capacity provided by EV in combined G2V and V2G Modes. Up/down capacity in combined G2V and V2G Mode is expressed as the sum of up/down capacity in G2V mode and V2G mode. Up capacity in G2V mode is the difference between the scheduled charging rate and the lower charging rate limit, whereas in V2G mode it is represented as the difference between the upper discharging rate limit and scheduled discharging rate. Similarly, down capacity in G2V mode is the difference between a minimum of maximum charging rate limit & an upper charging rate limit and a scheduled charging rate. Down capacity in V2G Mode is the difference between scheduled discharging rate and lower discharging rate limit.

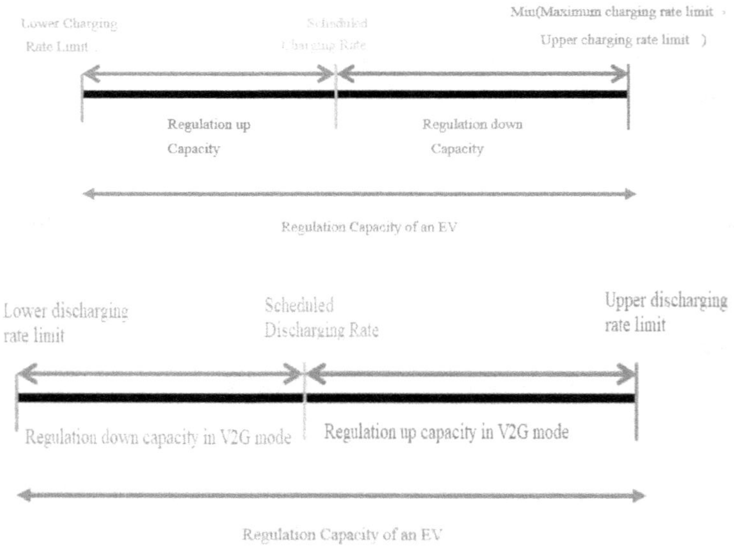

Fig. 7.2 Regulation capacity provided by an EV in G2V & V2G Modes

7.2.1 Operational Management Algorithm of EVA for Grid Integrated EVs

To enhance grid reliability, an efficient and scalable control algorithm is necessitated by EVA to monitor the operational assessment of EVs. For this, congestion management/transformer delivery capacity constraint compares the charging load of connected EVs against the rating of the transformer. The mismatch between demand and transformer rated capacity indicates the lacuna of allocation of energy and regulation, leading to sag/swell in the system. As a consequence of this, critical loads are affected as well. To overcome this issue, EVA performs charging of flexible EVs during off-peak times (surplus generation) and discharging during peak times (deficit generation) as shown in Fig.7.3.

If the charging demand of EVs is higher than the transformer rating, and there is no appropriate power allocation method, the transformer voltage will collapse, causing problems for other loads supplied by the same transformer. Thus, promising and effective power management and allocation algorithm are required to ensure that total power demand will not exceed the safe operating limit of the transformer. When generation is surplus than consumption, EVs in the form of dynamic/mobile energy storage absorb the surplus energy and deliver it back in case of consumption higher than a generation. EVs in the form of the controllable load is managed by EVA to support transformer load management.

EVA's algorithm reduces/increases $POP_{i,t}^{ch}$ / $POP_{i,t}^{dch}$ in case of deficit generation to reduce charging load and satisfy operational limits. Based on the baseload profile, EVA performs scheduling of charging and discharging of EVs. For example, if the base load is sufficiently low/high, EVs charge/discharge. EVs are having SOC close to the maximum value of SOC; participate in maintaining the reliability of the system by discharging to supply non-flexible loads.

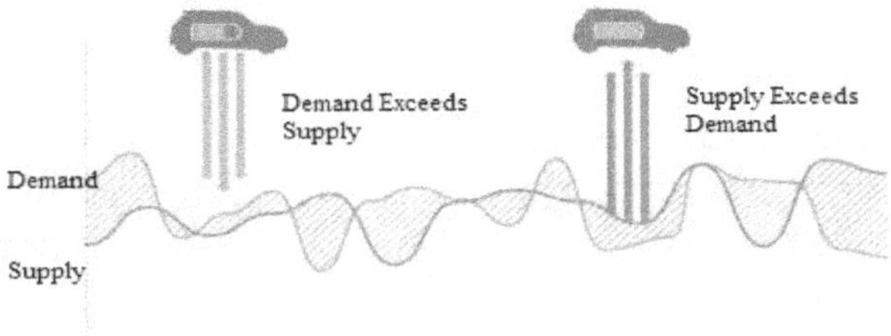

Fig. 7.3 The concept of V2G for Operational Management Algorithm of Grid Integrated EVs

7.2.2 Battery Degradation Cost Modelling

By modifying the degradation mechanisms as much as possible, the battery performance and life cycle can be extended; by proper management of the battery during storage and use during charging and discharging. Battery health management functions include battery balance, charge/discharge control, and temperature control. If the charge/discharge rate is high, battery degradation is high. Battery degradation causes calendar ageing and cycling ageing. Calendar ageing is influenced by temperature, time, and SOC. Cycling ageing is influenced by cycle number, charge/discharge rate and DOD.

$$DC_{i,t}^{\deg} = \frac{C^{bat} \cdot C_{Eb} + C_{labour}}{L_{cycles} \cdot C_{Eb} \cdot DOD} \tag{7.1}$$

$$Deg_{s,i,t} = DC_{i,t}^{\deg} \cdot \frac{POP_{s,i,t}^{dch} \cdot u_{s,i,t}^{dch}}{Efi^{dch}} \tag{7.2}$$

$$Cost_{s}^{\deg} = \sum_{i=1}^{N} \sum_{t=1}^{T} Deg_{s,i,t} \tag{7.3}$$

The battery degradation cost is as depicted by Eqn. (7.1) is a function of battery energy capacity cost C_{Eb} , the battery cost per kilowatt-hour C^{bat} , the battery replacement labour cost C_{labour} and the battery cycle life L_{cycles} at a specific depth of discharge (DOD) [52, 161-162]. Hence, the costs for an EVA are purchase costs of charging energy requirement from EV owners and the battery degradation cost paid to them. The cost of battery degradation is due to discharging electrical energy which is envisaged implementing equation (7.2) from V2G operation of typical EV and at any time t whereas (7.3) provides aggregated battery degradation cost over scheduling horizon.

Adopting TOU design methodology in Chapter 4, EVA specifies TOU rates to EV owners as shown in Table 7.1.

Table 7.1: Specified Electricity Rates to EV Owners designed by EVA from Forecasted RTP

	Off-Peak	**Shoulder Off-Peak**	**Flat**	**Shoulder On-Peak**	**On-Peak**
Duration	t5-t8	t2, t4	t1, t9	t3, t10, t24	t11-t22
TOU rates ($/kWh)	0.013	0.0156	0.017	0.019	0.023

7.2.3 Proposed Optimal Dispatch Algorithm of EVA Based on TOU prices and SOC Level of EVs

EVA adopts scheduled charging-discharging rates $POP_{s,i,t}^{ch} / POP_{s,i,t}^{dch}$ for operating modes of EVs according to their SOC level and time, from a fairness perspective to determine power dispatch to EV. When $SOC_{s,i,t}$ is close to its lower or upper limits, EVA sends a signal to change to idle mode. Opportune time is the appropriate time that is beneficial for all participants. If t belongs to off-peak duration ($t \in T^{OFF}$) and SOC is significantly lower than the maximum limit ($SOC << SOC^{\max}$), unidirectional power flow is performed as illustrated in Fig. 7.4. If t belongs to on-peak duration ($t \in T^{ON}$) and SOC is significantly higher than the minimum limit ($SOC >> SOC^{\min}$), EVA allows bidirectional power flow as embodied in Fig. 7.4.

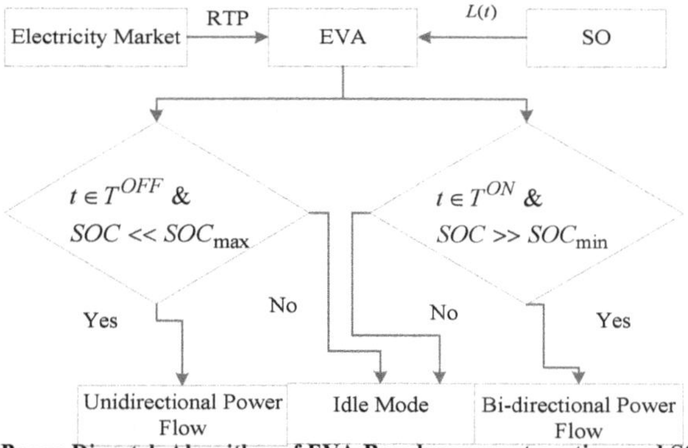

Fig. 7.4: Power Dispatch Algorithm of EVA Based on opportune time and SOC Level

7.2.4 Scenarios generation and reduction

The charging requirements of EVs are determined by random variables, such as their mobility, in terms of daily travelled distances and the EV owner driving profiles. For capturing baseline EV load, the efficient energy requirements of an EV fleet operation mobility behaviour of the EVs need to be modelled taking into account constant as well as stochastic parameters defining EVs operation viz. EV penetration level, charging level, availability, efficiency, charging strategy, type of EV, daily travelled distance and EV connection hours. To simulate the behaviour of EVs, arrival time and departure time are predicted from the history of the mobility pattern of ith EV [93]. It is assumed that they follow the Gaussian probability distribution function (pdf) as given below. Here,

μ_{ta_i} and μ_{td_i} are the mean, and σ_{ta_i}, σ_{td_i} are the standard deviation of the normally distributed ta_i and td_i, respectively.

$$pdf(\mathrm{ta}_i) = \frac{1}{\sigma_{ta_i}\sqrt{2\pi}} \exp\left(\frac{-\left(ta_i - \mu_{ta_i}\right)^2}{2\left(\sigma_{ta_i}\right)^2} \right)$$

$$pdf(\mathrm{td}_i) = \frac{1}{\sigma_{td_i}\sqrt{2\pi}} \exp\left(\frac{-\left(td_i - \mu_{td_i}\right)^2}{2\left(\sigma_{td_i}\right)^2} \right)$$

Fig. 7.5 shows the historic energy prices in the CAPITL zone of the NYISO market. Monte Carlo Simulation has been used to simulate scenarios of uncertain variables accurately parameters with a normal distribution *viz.* energy and regulation market prices and mobility behaviour (times of arrival and departure). In the proposed work, KD backward scenario reduction approach is used to reduce the generated scenarios [126]. KD assures maximum probable scenarios without violating necessary tolerance criteria. The KD matrix scenario reduction equation is given below.

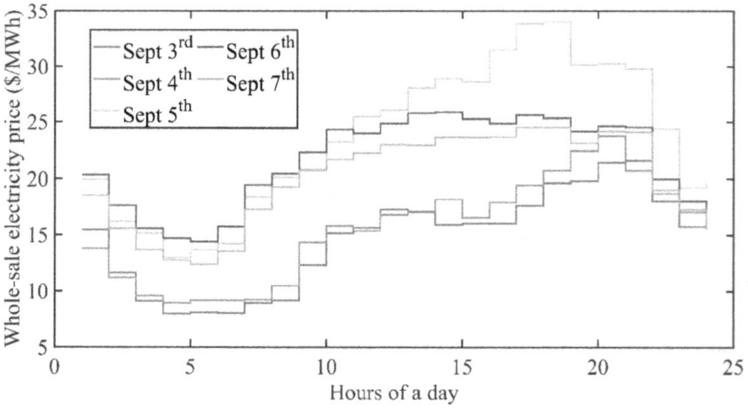

Fig. 7.5: Daily electricity wholesale prices in the CAPITL zone of New York ISO market.

$$KD\left(\xi,\xi'\right) = \sum_{\xi\in\Omega/\Omega_s} \gamma_{\xi} \cdot \upsilon\left(\xi,\xi'\right)$$

$$\upsilon\left(\xi,\xi'\right) = \left\| P_{\xi} - P_{\xi'} \right\|, \forall \xi,\xi' \in \Omega,$$
$$\Omega = 1,2,\cdots,N_{\xi}$$

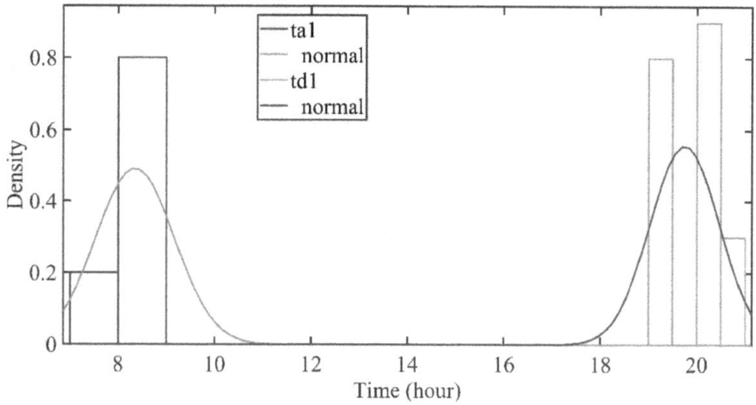

Fig. 7.6: Density & normal fit for forecasted times of arrival and departure for the first EV

P_{ξ} and $P_{\xi'}$ are the individual probability. 1000 scenarios of prices are generated for 24-hour with Monte Carlo Simulation. Market price uncertainty modelling is obtained using a normal probability distribution factor. The forecasted energy and regulation market price data from reference [134] is taken as mean 19.86375 and 19.59583 with SD of 4.164701 and 4.07938 to represent volatility. By using the KD matrix, 20 scenarios are obtained from the 1000 generated scenarios. Both reduced and generated scenarios have a 95% confidence interval. E.g. Fig. 7.6 indicates the density and normal fit for forecasted times of arrival and departure for the first EV.

7.2.5 Risk Management

Risk management relating to the V2G scheduling of EVA in an uncertain environment could be possible by including an additional term called risk measure, which measures the risk affecting the profit distribution in the objective function. Diverse risk managing methods have been anticipated in the existing literature [90] for scheduling conventional market partakers such as distribution companies, generation companies, large consumers and load-serving entities.

Risk measures extensively used for characterizing risk are the Value at Risk (VaR) and the Conditional value at risk ($CVaR$). These measures require the use of a probability, ($1-\alpha$), which refers to the part of the profit distribution that is desired to be managed to hedge the risk. The VaR is the ($1-\alpha$)-quantile of the profit distribution, whereas the $CVaR$ is the expected value of the profit distribution below that ($1-\alpha$)-

quantile. Since these measures do not use a profit target value, they are more suitable for modelling risk. Also, the *CVaR* is a coherent risk measure. Currently, the *CVaR* is widely used because, besides being a coherent risk measure, it can be expressed using a linear formulation. For these two reasons, the *CVaR* is the most used risk measure in problems of electricity markets.

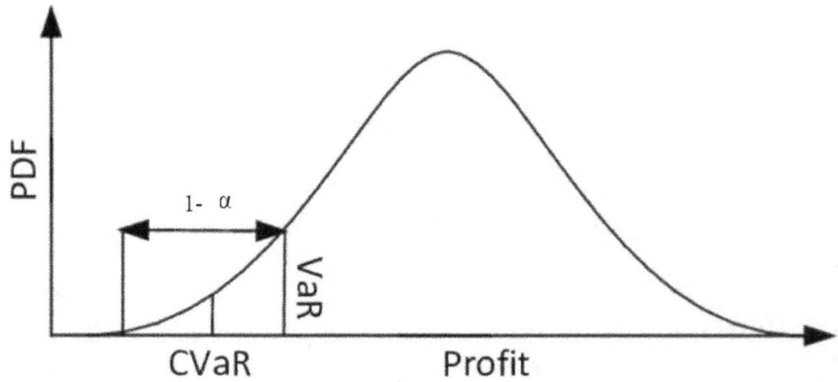

Fig. 7.7: Illustrating the concepts of *VaR* and *CVaR*

VaR is a statistic that measures and quantifies the level of financial risk within a firm, portfolio or position over a specific time frame. One can apply *VaR* calculations to specific positions or whole portfolios or to measure firm-wide risk exposure. Under the Monte Carlo method, *VaR* is calculated by randomly creating many scenarios for future rates using non-linear pricing models to estimate the change in value for each scenario, and then calculating the *VaR* according to the worst losses. *VaR* is a measure of market risk that objectively combine the sensitivity of the portfolio to market changes and the probability of a given market change. *VaR* calculates the maximum loss expected (or worst-case scenario) on an investment, over a given period and given a specified degree of confidence. *CVaR* is an improved version of *VaR* which is well-defined as the difference between the expected profit and the average of potential profit values which are less than *VaR*. Therefore, the *CVaR* is also known as mean excess loss or average *VaR*. *CVaR* is a coherent risk measure satisfying all four properties i.e., monotonicity, subadditivity, positive homogeneity, and translation invariance [90]. Fig. 7.7 illustrates the *VaR* and *CVaR* values using the probability distribution function of a hypothetical profit. $CVaR$ represents the mean of the profit of the scenarios with the return less than the VaR value.

7.3 Risk-averse EVA's V2G Scheduling Problem formulation

The maximum charging rate limit $P_{s,i,t}^{m\max}$ is the power required to charge the ith EV to its desired SOC in one-time slot t which is considered by the subsequent equation (7.4). Here, $SOC_{s,i,t}$ is the SOC level, BC_i is the battery capacity (kWh) and Ef_i is the charging efficiency of ith EV battery.

$$P_{s,i,t}^{m\max} = \left(SOC_i^{fin} - SOC_{s,i,t}\right)\cdot h_{i,t}\cdot\frac{BCi}{Efi} \quad \forall s,i,t \tag{7.4}$$

The vehicle battery discharge and charge limits $P_i^{\max dch}, P_i^{\max ch}$ considering, respectively, the battery state of charge and the battery capacity and the previous period stored energy are given as follows:

$$\frac{1}{Ef_i^{dch}}\cdot POP_{i,t}^{dch} \le SOC_{i,t-1} \quad \forall t\in\{1,\cdots,T\};\ \forall i\in\{1,\cdots,N\};$$

$$Ef_i^{ch}\cdot POP_{i,t}^{ch} \le \frac{\rho_i^{\max}}{BC_i} - SOC_{i,t-1} \quad \forall t\in\{1,\cdots,T\};\ \forall i\in\{1,\cdots,N\};$$

To analyse the effects of β on the EVA's scheduling strategy, three numerical cases are considered as summarised in Table 7.2, showing the validity of the proposed approach for V2G paradigm managing. In the first case of RNDM neglecting mobility behaviour uncertainty, β=0, therefore risk measure expression $CVaR$ is not considered within the objective. However, case 2 of RADM ignores behaviour uncertainty rather takes into account the risk measure term $CVaR$. EVA is hedged against risks due to risk control. The last case 3 of RNDM and RADM assesses the impact of EV owners' mobility behaviour uncertainty on V2G scheduling of EVA.

Table 7.2: Summary of numerical cases

Numerical cases	Market price uncertainty	Mobility behaviour uncertainty	Risk Management
Case1	*	x	x
Case2	*	x	*
Case3	*	*	*

In simplified form equations (7.5) and (7.6) show the charge/discharge limit for each EV considering the battery charge/discharge rate is given above where, $P_{i,\max}^{dch}$ / $P_{i,\min}^{dch}$ and $P_{i,\max}^{ch}$ / $P_{i,\min}^{ch}$ are the maximum or minimum power discharge and charge of EV i. In constraint (7.7), a value is assigned to the SOC of each plugged-in EV at every timeslot. As per (7.8), SOC must lie within lower and upper limits. (7.9) ensures in a specific scenario, i^{th} EV either charges or discharges but not both at a particular timeslot.

$$P_{i,\min}^{ch} \cdot u_{s,i,t}^{ch} \le POP_{s,i,t}^{ch} \le P_{i,\max}^{ch} \cdot u_{s,i,t}^{ch} \quad \forall s,i,t \tag{7.5}$$

$$P_{i,\min}^{dch} \cdot u_{s,i,t}^{dch} \le POP_{s,i,t}^{dch} \le P_{i,\max}^{dch} \cdot u_{s,i,t}^{dch} \quad \forall s,i,t \tag{7.6}$$

The SOC represents the stored energy in the battery of the vehicle i at the end of the period t. The SOC updating is according to the SOC that remained from the previous period and the charge/discharge in that period as follows (7.7): where Ef_i^{ch} and Ef_i^{dch} represent, respectively, the G2V charging and V2G discharging efficiency coefficients of EV i.

$$SOC_{s,i,t} = \begin{cases} SOC_{s,i,t-1} + Ef_i^{ch} \cdot \dfrac{POP_{s,i,t-1}^{ch}}{BatCapi} - \dfrac{1}{Ef_i^{dch}} \cdot \dfrac{POP_{s,i,t-1}^{dch}}{BatCapi}, & ta_i < t < td_i \\ SOC_i^{ini}, & t = ta_i \\ SOC_i^{des}, & t = td_i \end{cases} \tag{7.7}$$

Depletion of EV battery up to a certain minimum level $\rho_i^{\min}$ and charging up to a maximum level $\rho_i^{\max}$ are ensured by following equations to prevent loss of battery life [167].

$$SOC_{i,t} \le \frac{\rho_i^{\max}}{BC_i} \quad \forall t \in \{1,\cdots,T\}; \ \forall \mathrm{i} \in \{1,\cdots,N\};$$

$$SOC_{i,t} \ge \frac{\rho_i^{\min}}{BC_i} \quad \forall t \in \{1,\cdots,T\}; \ \forall \mathrm{i} \in \{1,\cdots,N\};$$

where $\rho_i^{\min}$ and $\rho_i^{\max}$ is defined based on the battery capacity limit for each EV that is calculated as follows:

$$\rho_i^{\max} = BC_i \cdot SOC_i^{\max} \ \forall i \in \{1,\cdots,N\}$$

$$\rho_i^{\min} = BC_i \cdot SOC_i^{\min} \ \forall i \in \{1,\cdots,N\}$$

Where BC_i represents the battery capacity of EV $i; SOC_i^{\max}$ and $SOC_i^{\min}$ are, respectively, the maximum and minimum SOC specified by EV owners considering battery life. This can be simplified as (7.8) with SOC of EV lying within lower and upper limits in V2G operations as:

$$\mathrm{SOC}_{\mathrm{i,min}} \cdot h_{i,t-1} \le SOC_{s,i,t} \le \mathrm{SOC}_{\mathrm{i,max}} \cdot h_{i,t-1} \quad (7.8)$$

$$u_{s,i,t}^{ch} + u_{s,i,t}^{dch} \le 1 \cdot h_{i,t} \ \forall s,i,t; u_{s,i,t}^{ch}, u_{s,i,t}^{dch} \in \{0,1\} \quad (7.9)$$

Binary variables $u_{i,t}^{ch}$ and $u_{i,t}^{dch}$ in (7.9) indicates charging and discharging operation for value 1; both values 0 indicates the idle mode. Regulation up/down capacity is the sum of regulation up/down capacity provided in G2V and V2G mode as given by (7.10) and (7.11) respectively. Regulation up capacity provided in G2V mode is a difference between the scheduled charging rate and its lower limit. Regulation up capacity provided in V2G mode is the difference between the upper limit of discharging rate and scheduled discharging rate. Regulation down capacity provided in G2V mode is a difference of a minimum of maximum charging rate limit and upper limit of charging rate and scheduled charging rate. Regulation down capacity provided in V2G mode is the difference of scheduled discharging rate and lower limit of discharging rate. Total regulation capacity is the sum of the regulation up and down capacity as per (7.12).

$$uc_{s,i,t} = \left(POP_{s,i,t}^{ch} - P_{i,\min}^{ch} \cdot u_{s,i,t}^{ch}\right) \cdot h_{i,t} + \left(P_{i,\max}^{dch} \cdot u_{s,i,t}^{dch} - POP_{s,i,t}^{dch}\right) \cdot h_{i,t} \quad (7.10)$$

$$dc_{s,i,t} = \left(min\left(P_{s,i,t}^{m\max}, P_{i,\max}^{ch}\right) \cdot u_{s,i,t}^{ch} - POP_{s,i,t}^{ch}\right) \cdot h_{i,t} + \left(POP_{s,i,t}^{dch} - P_{i,\min}^{dch} \cdot u_{s,i,t}^{dch}\right) \cdot h_{i,t} \quad (7.11)$$

$$rc_{s,i,t} = uc_{s,i,t} + dc_{s,i,t} \quad (7.12)$$

$$Cost_{s,i} = \sum_{t=1}^{T} EP_{s,t}^{RTP} \cdot POP_{s,i,t}^{ch} \cdot h_{i,t} + Deg_{s,i,t} \cdot h_{i,t} \quad (7.13)$$

Charging cost for EVA is the sum of procurement cost and degradation cost as depicted by (7.13). The upper limit of EV owners' charging cost (CB) could be determined from the unregulated baseline charge scheduling case [65]. The revenue from charging must be less than this CB as expressed by (7.14).

$$\sum_{t=1}^{T}\left(Mk_t + EP_t^{TOU}\right)\cdot POP_{s,i,t}^{ch}\cdot h_{i,t} \le CB \tag{7.14}$$

Constraint (7.15) is the congestion management constraint preventing overloads of distribution transformer due to connected baseload, aggregated charging load and regulation down capacities at each timeslot is less than or equal to the sum of transformer delivery capacity and aggregated discharging resource and regulation up capacities at each timeslot.

$$\left| l_t^{base} + \sum_{s=1}^{N_s}\sum_{i=1}^{N} POP_{s,i,t}^{ch}\cdot h_{i,t} + \sum_{s=1}^{N_s}\sum_{i=1}^{N} dc_{s,i,t}\cdot h_{i,t} - \sum_{s=1}^{N_s}\sum_{i=1}^{N} POP_{s,i,t}^{dch}\cdot h_{i,t} - \sum_{s=1}^{N_s}\sum_{i=1}^{N} uc_{s,i,t}\cdot h_{i,t} \right| \le TDC \tag{7.15}$$

Revenue of EVA comes from regulation services to SO, charging services to EV owners at TOU price and discharged energy in the market at RTP as illustrated from (7.16). Here, the first term of (7.16) indicates revenue due to total regulation capacity; the second term directs revenue due to charging of EVs; the third term shows revenue due to discharging of EVs at real-time energy prices. (7.17) expresses profit from s^{th} scenario and the ith EV as the difference of revenue and cost. Expected profit is the sum of the product of sth scenario probability and profit as in (7.18).

The objective of the optimal risk aversion decision-making (RADM) scheduling of EVA taking into account the, $CVaR$ is formulated as a stochastic programming problem through (7.19). The first term is the main profit objective and the second one is $CVaR$. The two terms respectively expected profit and $CVaR$ are trade-off through β. Here, while considering/ignoring mobility uncertainty, the probability of a particular connected EV i in the scenario s is depicted as $prob_{i,s} / prob_s$. (7.20-7.22) defines Conditional value at risk $CVaR$ as used in portfolio optimization for effective risk management. $CVaR$ is derived by taking a weighted average of the "extreme" losses in the tail of the distribution of possible returns, beyond the value at risk (VaR) cut-off point. Where ϕ_s, β, and α are respectively the auxiliary variable, risk aversion parameter, and confidence level.

$$Revenue_{s,i} = \sum_{t=1}^{T}\left(RP_{s,t}\cdot rc_{s,i,t} + \left(Mk_t + EP_t^{TOU}\right)\cdot POP_{s,i,t}^{ch} + EP_{s,t}^{RTP}\cdot POP_{s,i,t}^{dch}\right)\cdot h_{i,t} \tag{7.16}$$

$$Profit_{s,i} = \text{Re}venue_{s,i} - Cost_{s,i} \tag{7.17}$$

$$Expected\ profit = \sum_{s=1}^{N_s} prob_s \cdot \sum_{i=1}^{N} \Pr ofit_{s,i} \tag{7.18}$$

$$\max_{\text{POP}^{ch}_{s,i,t},\text{POP}^{dch}_{s,i,t},\text{rc}_{s,i,t},\phi_s,\eta} \quad RADM = (1-\beta)\cdot Expected\Pr ofit + \beta \cdot CVaR \tag{7.19}$$

$$CVaR = \eta - \frac{1}{(1-\alpha)} \cdot \sum_{s=1}^{N_s} \Pr ob_s \cdot \phi_s \tag{7.20}$$

$$\eta - \sum_{i=1}^{N} \Pr ofit_{s,i} \le \phi_s \tag{7.21}$$

$$\phi_s \ge 0 \tag{7.22}$$

7.4 Simulation Results

Proposed TOU pricing based V2G scheduling and dispatch of EVA to support grid stability and reduce charging costs, utilizing historical data from NYISO market information [134]. The average of weekday electricity prices (September 2018) is used to predict the TOU prices. To make a low computational burden, at first, 1000 price scenarios are produced and assumed that the market prices follow normal distribution function, then the produced scenarios reduced to 20 representative scenarios using Kantorovich distances and backward reduction method.

The price scenarios are shown in Fig. 7.8 and 7.9. Mobility behaviour is modelled using NHTS data [93, 136]. Considering scenarios mentioned above, the presented optimization, including equations (7.1)-(7.18), and (7.20)-(7.22) as a risk-neutral and risk-averse problem is modelled as mixed-integer non-linear programming (MINLP) and solved by GAMS optimization software using DICOPT solver [137]. In the following, the results are shown and discussed for the risk-neutral and risk-averse cases with and without mobility behaviour uncertainty.

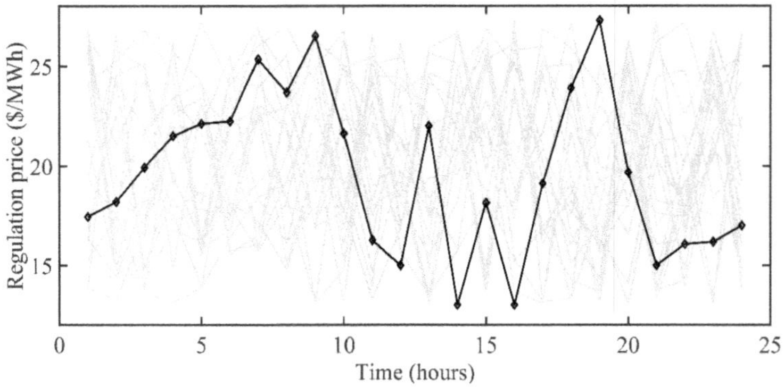

Fig. 7.8: Reduced scenarios of the regulation price (green) along with the forecasted price (black)

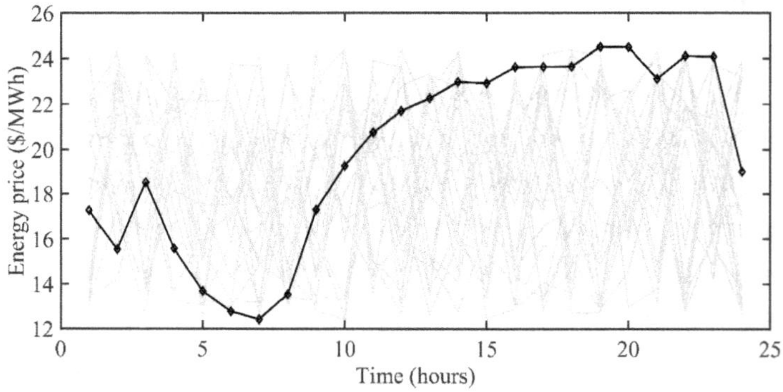

Fig. 7.9: Reduced scenarios of energy price (green) along with the forecasted price (black)

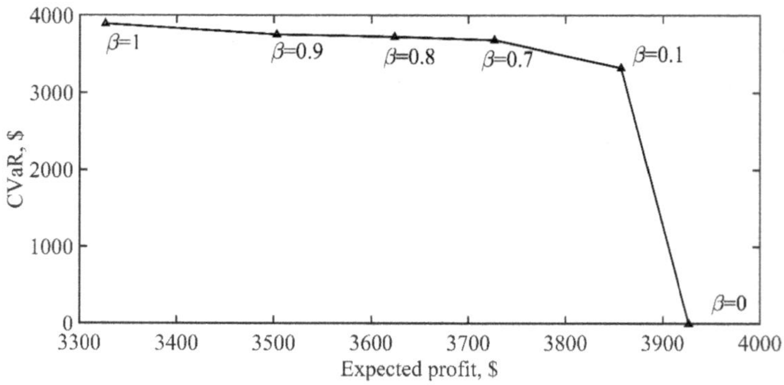

Fig. 7.10: The efficient frontier of *CVaR* versus expected profit for different risk appetite

Value of *CVaR* increases as the value of weight parameter β increases as shown in Fig. 7.10. Fig.7.10 illustrates a combination of efficient points of *CVaR* and expected profits at distinct values of risk aversion parameter β, termed as an efficient frontier.

When β raises, expected profit becomes less significant concerning risk. β is decided to depend upon EVA's appetite. A risk-averse EVA selects a large value for β and vice versa. As β increases, the expected profit decreases. At β=1, EVA earns a minimum expected profit. The expected profit for β=1 as compared to β=0.1 decreases by 13.74% to obtain an increment of 5.2% in $CVaR$. EVA utilizes this efficient frontier as an instrument to resolve the trade-off between the expected profit and risk.

Response to TOU prices from different price-sensitive EV owners leads to an efficient frontier with lower $CVaR$ corresponding to higher expected profit. An increase in the charging cost impacts participation of EV owners and therefore, EVA's profit. EVA attempts to manage risk on account of indeterminacy of the energy market through regulation market participation and vice versa. In EVs' charging mode, the selection of higher β values upsurges $CVaR$ along with a decrease of expected profit. As β increases, procurement/sale in the energy market reduces/increases indicating a decrease in charging rate and increase in discharging rate thereby peak shaving supports SO during peak hours. Therefore, provided regulation up capacity decreases and regulation down capacity increases. Similarly, if β decreases, procurement/sale in the energy market increases/reduces indicating an increase in charging rate and decrease in discharging rate thereby supporting valley filling during off-peak hours. Therefore, provided regulation up capacity increases and regulation down capacity reduces. So, the EVA tries to increase participation in the regulation market to prevent a significant reduction in expected profit.

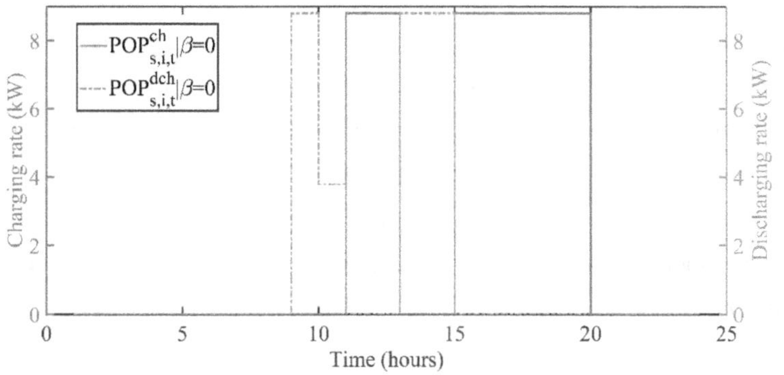

Fig. 7.11: Charging/Discharging rate for a typical scenario1 and EV1 at time t with β=0

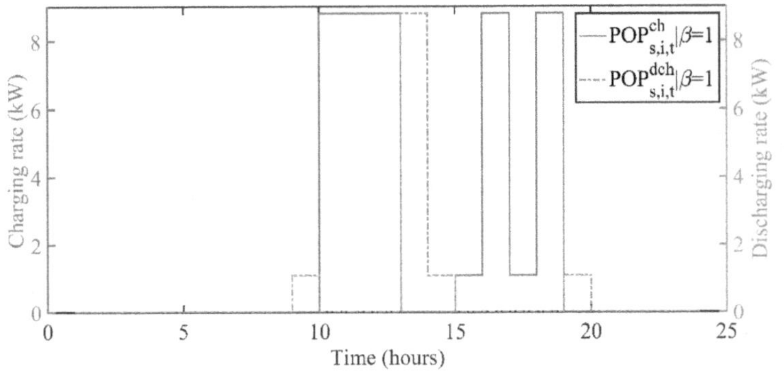

Fig. 7.12: Charging/Discharging rate for a typical scenario1 and EV1 at time t with β=1

Figs. 7.11 and 7.12 indicate charging/discharging rate for a typical scenario1 and EV1 at time t for varying risk aversion parameters from 0 to 1. For risk-neutral case, the charging/discharging rate is not distributed in nature, therefore EVA's expected profit is high due to increased revenue from charging and regulation services. However, a high value of charging/discharging rate may cause a generation-demand imbalance in the system. Or in other words, it raises instability issues and degrading EV batteries. Risk-averse case shows that charging rate is higher in off-peak duration and lower in on-peak duration while discharging rate is higher in on-peak duration and lower in off-peak duration to optimize benefits for both SO and EV owners. SO is satisfied by avoiding charging at peak times thereby avoiding costly generation units and system overloading issue is mitigated.

Figs 7.13 & 7.14 represents aggregated charging load variation respectively according to baseload and TOU price at an office parking lot for different risk appetite indicating that for lower baseload/higher TOU price charging rate for the risk-averse case is lower as compared to risk-neutral case based on connection hours. Fig. 7.15 represents power injected to the grid according to baseload at an office parking lot for different risk appetite indicating that for higher/lower baseload power injected for the risk-averse case is higher/lower as compared to risk-neutral case. Fig. 7.16 indicates SOC variation of a particular EV operating in G2V/V2G modes at different timeslots t with 80% SOC at departure.

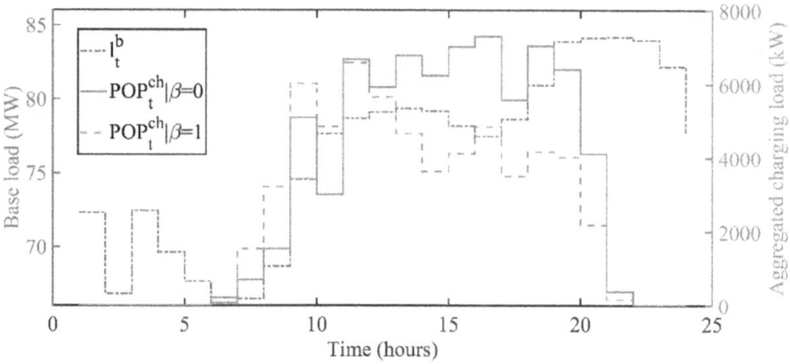

Fig. 7.13: Aggregated charging load variation according to baseload at an office parking lot for different risk appetite

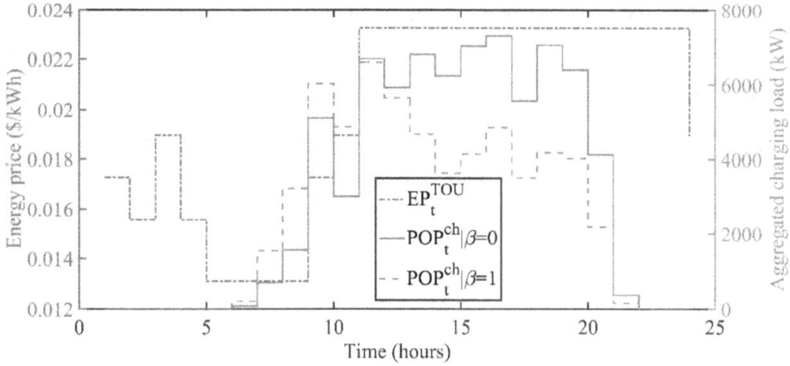

Fig. 7.14: Aggregated charging load variation according to TOU price at an office parking lot for different risk aversion parameters

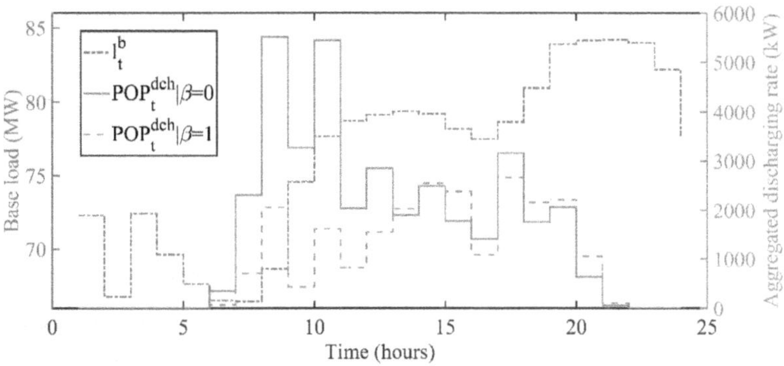

Fig. 7.15: Power injected to the grid in V2G mode according to baseload at an office parking lot and different timeslot t for varying risk appetite

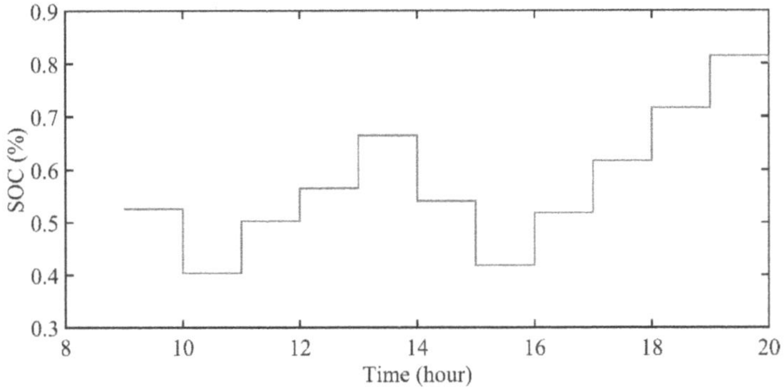

Fig. 7.16: SOC variation of a particular EV operating in G2V/V2G modes at different timeslot t

Figs. 7.17 and 7.18 indicate regulation up and down capacities provided by risk-neutral and risk-averse EVAs. As it is clear that for the risk-neutral case from t_6-t_{10} (off-peak hours), regulation up/down capacity is lower/higher as compared to risk-averse case. From t_{11}-t_{21} (peak hours), risk-neutral up/down capacity is higher/lower as compared to risk-averse case. This is due to the price-responsive behaviour of EV owners, supporting valley filling for off-peak hours and peak shaving at peak hours respectively.

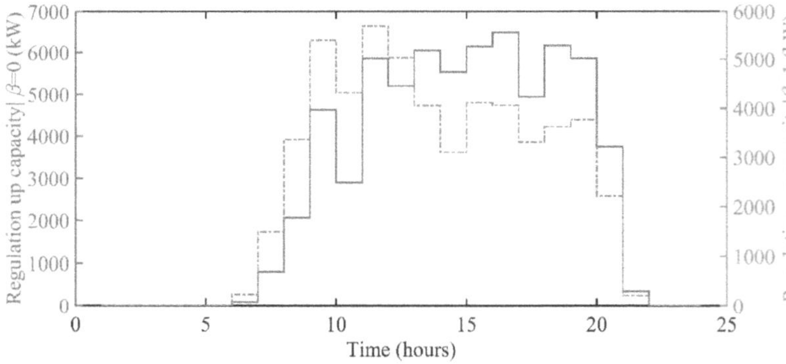

Fig. 7.17: Regulation up capacity offered by EVA at time t according to different β

The value of the profit for each scenario relating to the risk-neutral case is plotted in Fig. 7.19. The resulting expected profit is \$3924.176. It is observed that the profit exhibits high volatility varying from \$3670.862 to \$4133.537. Since the risk-neutral problem only focuses on maximizing the expected profit, the risk of experiencing low profits in some scenarios is not avoided. Observe that there exists a probability equal to 0.45 of experiencing lower profits than expected value.

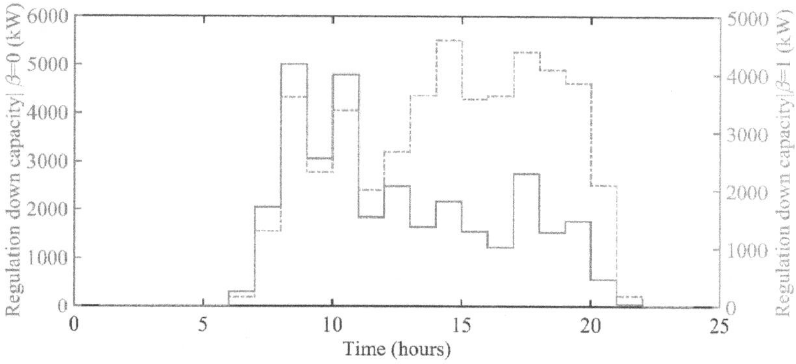

Fig. 7.18: Regulation down capacity offered by EVA at time t according to different β

For the risk-averse situation, the expected profit is equal to \$3326.81, that is, a value much smaller than that obtained for the case $\beta = 0$ (\$3924.176). This result indicates that the regulation market is an effective tool to reduce the risk associated with profit variability. In Figs.7.19 & 7.20, the cumulative distribution function (CDF) of profit considering different risk-aversion parameters 0 and 1 have been plotted. As can be seen, as the V2G scheduling strategy is changed to risk-averse, the curves submitted to the market is altered. From Figs., the CDF of the profit under the proposed integrated DR and risk-averse V2G problem dominates the CDF of β=0 and would be chosen by a risk-averse decision-maker EVA. Moreover, curve 2 is more restrictive than curve1. As can be seen, for a particular probability in both cases, the gained profit is lower than the profit of risk-neutral case and is more conservative.

Fig. 7.21 indicates the variation of expected cost and revenue with varying β, validating reduction in both cost and revenue due to reduced procurement/sale of energy from lower charging/discharging rate as per battery health point of view. Sensitivity analysis of performance metrics of EVA w.r.t battery degradation cost for different β is performed. The obtained results in Fig. 7.22 and 7.23 show that according to EV battery capacity and by considering degradation cost, the profit of the risk-neutral and risk-averse EVA is decreased by 33.786% and 27.9178% in comparison with a case without considering degradation cost. It is worth mentioning that the obtained results are sharply sensitive to market price scenarios and degradation cost modelling parameters, moreover, multi-market participation is profitable for EVA, only when there is a significant gap between off-peak and peak time prices, so that the revenues cover the payment costs.

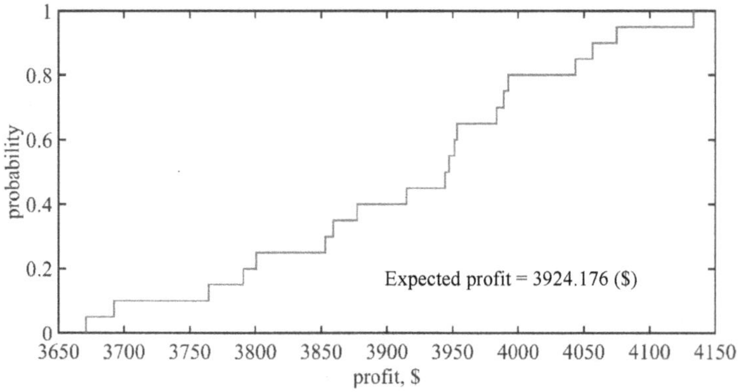

Fig. 7.19: The cumulative profit distribution function for β=0

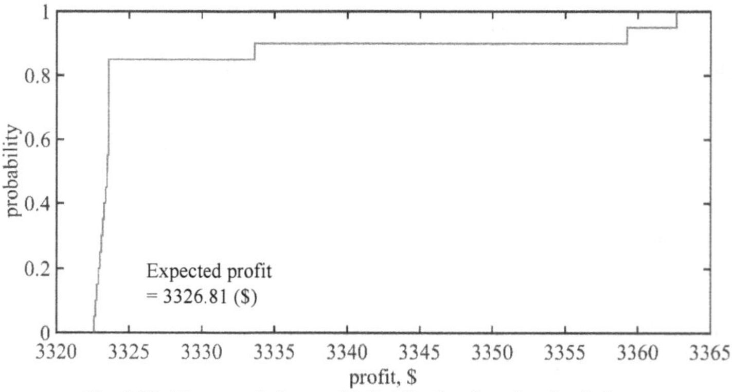

Fig. 7.20: The cumulative profit distribution function for β=1

Moreover, Figs. 7.22 and 7.23 show that for lower costs of battery replacement, revenue or profit is sensitive but with higher replacement costs the increased V2G connected capacity compensates the cost through increased earned revenues. Thus, the variation of profit and revenue saturates while cost reduces. Or in other words, V2G becomes independent of battery degradation due to TOU-PBDR. V2G energy arbitrage and regulation up capacity sold to SO reduce as battery replacement cost increases. For higher battery replacement/degradation costs, dispatch of regulation down capacity is higher as compared to dispatch of regulation-up capacity. For lower replacement costs, peak shaving is preferred and dispatch of regulation-up is performed. Thus, the proposed method could reduce the charging load and also support by discharging in peak hours followed by more load recovery in off-peak hours. Henceforth, peak shaving and valley filling improve the economic efficiency of the power system.

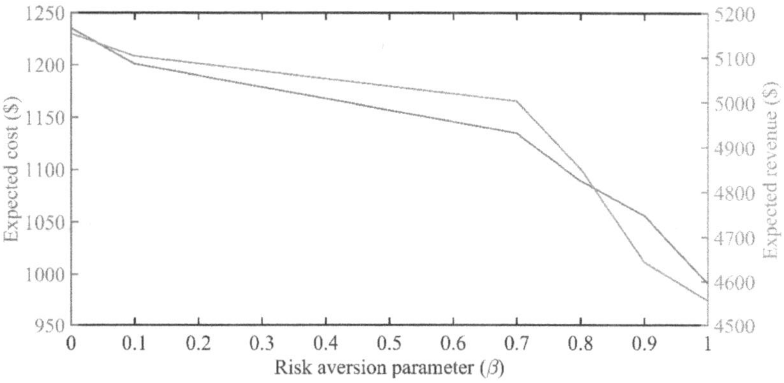

Fig. 7.21: Performance parameters of EVA according to different β

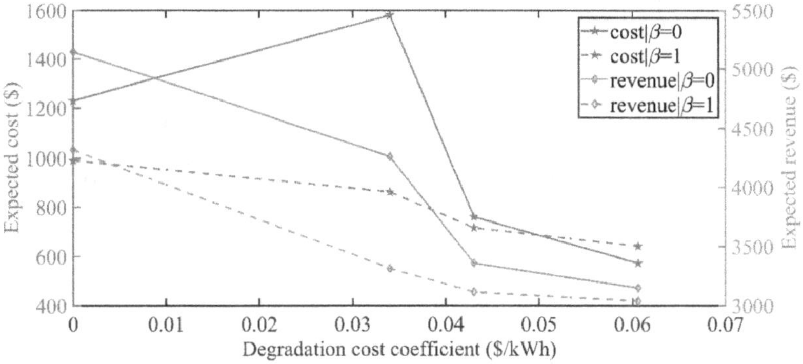

Fig. 7.22: The sensitivity of expected cost and revenue concerning degradation cost coefficient

Sensitivity analysis of economic objectives/performance metrics of EVA w.r.t Round Trip Efficiency RTE (Ratio of the energy retrieved from the battery to the energy provided to the battery, when coming back to the same SOC state) for different β is performed. It is observed that these metrics improve as efficiency enhances. As EVA becomes risk-averse, its profit reduces as compared to risk-neutral case. Figs. 7.24 and 7.25 illustrate the sensitivity analysis of cost, revenue and profit concerning RTE. Profit sensitivity is lower as compared to variations in cost and revenues.

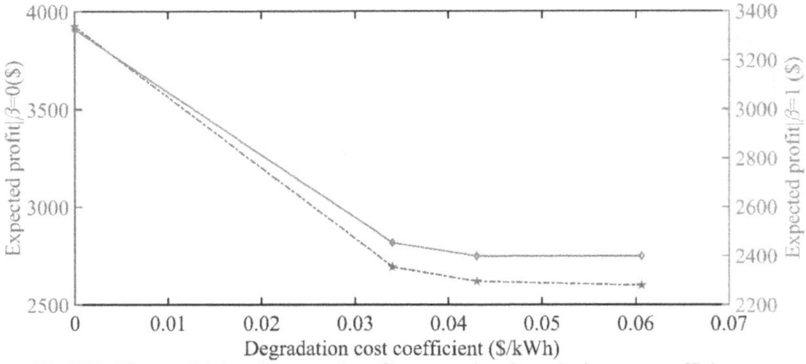

Fig. 7.23: The sensitivity of expected profit concerning degradation cost coefficient

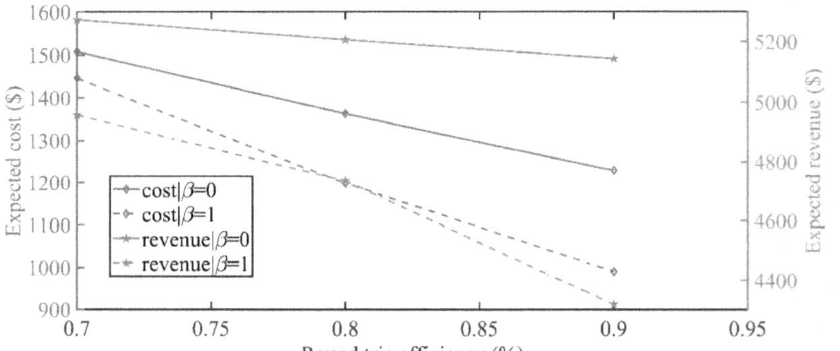

Fig. 7.24: The sensitivity of expected cost and revenue w.r.t. round trip efficiency

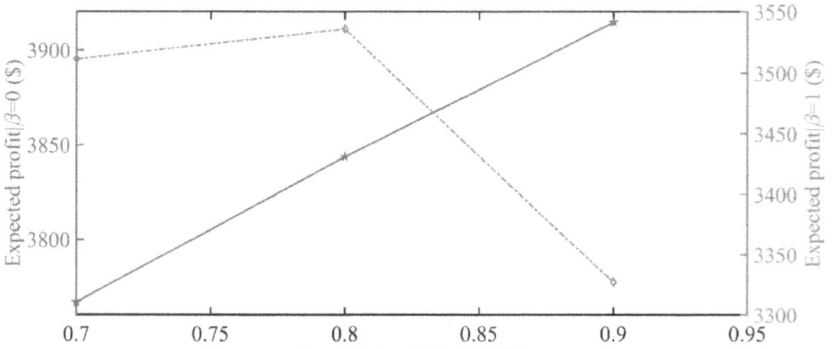

Fig. 7.25: The sensitivity of expected profit for round trip efficiency

Tables 7.3 and 7.4 provides a comparison between risk-neutral and risk-averse problems for 1000 EVs indicating a decrease in value of expected profit of risk-averse problem, as β increases. A decrement in profit confirms that considering risk in the scheduling process will lead to a lower value of profit for EVA.

Table 7.3 Simulation results of Risk-neutral Case for 1000 evs

Expected revenue ($)	Expected cost ($)	Expected Profit ($)
5143.951	1229.34	3924.176

Table 7.4 Simulation results of Risk-Averse Case for 1000EVs

Expected revenue ($)	Expected cost ($)	Expected profit ($)
4556.771	1056.247	3502.909

In Table 7.5, the expected cost of V2G increases/decreases in the case of utility/designed TOU while it decreases in the case of G2V. EVA's expected profit in case of utility/designed TOU decreases significantly/negligibly in G2V mode. Whereas interesting thing to note is that utility/designed PBDR decreases/increases expected profit in V2G mode. Thus, it can be concluded from Tables 7.5 and 7.6 that the execution of designed TOU-PBDR is beneficial as compared to RTP-PBDR for both the involved entities: EVA and EV owners. EVA's expected profit is more in designed TOU due to shifting. The system parameters utilized in the V2G scheduling formulation are specified in Table 7.7.

Table 7.5 Sensitivity of performance parameters with PBDR type for case2 for utility (Con Edison) TOU to EVA

	V2G		G2V	
	RTP-PBDR	Utility TOU-PBDR	RTP-PBDR	Utility TOU-PBDR
Expected Cost	1232.6	1288.58	1040.88	957.782
$CVaR$	0	0	1382.29	1083.72
Expected profit	3915.8	3905.71	1650.94	1477.13
Expected Revenue	5140.3	5184.48	2506.23	2179.997

From Tables 7.8 and 7.9, comparing the expected profit in V2G mode decreases significantly/negligibly obtained without/with mobility behaviour uncertainty. In G2V mode, it reduces negligibly/significantly obtained without/with mobility behaviour uncertainty.

The driving behaviour of EVs affects available capacity, SOC and potential profit of EVA. System parameters utilized in the V2G scheduling formulation are specified in Table II. The problem is modelled using MINLP and solved using DICOPT (DIscrete and Continuous OPTimizer) and SBB (Simple Branch and Bound) solver through GAMS software [137]. EVs with higher SOC will participate in the regulation market and discharge by V2G mode. The sale of energy and regulation capacity both increases the revenue of EVA. EVs with lower SOC will charge in G2V mode. EVA earns not only through the sale of charging energy to EV owners for the off-peak duration but also by injecting discharged energy to the grid during peak hours. Based on price scenarios from forecasted market price and baseload profile from SO, EVA provides energy and regulation services to SO through TOU-PBDR of EV owners.

To exploit V2G services, regulation capacity is transacted more as compared to energy since this service is more commendable due to higher regulation prices. For a win-win situation for all involved entities, most of the charging is performed in off-peak duration and discharging in peak duration to alleviate pressure on the SO.

Table 7.6 Sensitivity of performance parameters with PBDR type for case2 for designed TOU from EVA

	V2G		G2V	
	RTP-PBDR	**Designed TOU-PBDR**	**RTP-PBDR**	**Designed TOU-PBDR**
Expected Cost	1232.62	1229.338	1040.88	991.0638
$CVaR$	0	0	1382.29	1328.717
Expected profit	3915.84	3924.176	1650.94	1642.757
Expected Revenue	5140.38	5143.951	2506.23	2398.701

Table 7.7: System Parameters Used for V2G scheduling of EVA

EV Specifications	Value	EV Specifications	Value
Battery Capacity	80 kWh	Mean of departure time	7 pm
Minimum SOC	10%	The standard deviation of arrival time	2hrs
Maximum SOC	90%	The standard deviation of the departure time	2hrs
Charging/discharging efficiency	0.9	Desired SOC of each EV	0.9
Mark-up price	40% of Energy price	The lower limit of the charging/discharging rate	1.1kW
Mean of arrival time	7 am	The upper limit of the charging/discharging rate	8.8kW

Table 7.8: Expected Cost, revenue, and profit without mobility behaviour uncertainty for respectively RNDM in case1and RADM in case2

	V2G		G2V	
	RNDM $\beta=0$	RADM $\beta=1$	RNDM $\beta=0$	RADM $\beta=1$
Expected cost (\$)	1229.3	990.8014	1110.12	911.1167
Expected Revenue (\$)	5143.9	4317.77	2548.936	2400.782
Expected profit (\$)	3924.1	3326.81	1674.514	1604.688

Table 7.9: Expected Cost, revenue, and profit with mobility behaviour uncertainty for case3

	V2G		G2V	
	RNDM $\beta=0$	RADM $\beta=1$	RNDM $\beta=0$	RADM $\beta=1$
Expected cost (\$)	1208.607	1209.224	1087.92	1090.801
Expected Revenue (\$)	5227.074	5226.807	2624.304	2678.311
Expected profit (\$)	3932.086	3931.835	1846.831	1439.77

Fig. 7.26 represents different TOU structures from utility and designed from EVA. Figs. 7.27 and 7.28 show the sensitivity to cost and revenue of EVA with a varying number of EVs in V2G and G2V modes respectively. For a smaller group, cost and revenue are lower. For a larger group, the cost is significantly lower, and revenue is considerably higher. There is a 9.31% reduction in cost and a 7.90% reduction in revenue from risk-neutral to risk-averse in V2G mode as compared to a 0.068% increase in cost and 0.8983% reduction in revenue from risk-neutral to risk-averse in G2V mode for 1000 EVs.

Figs. 7.29 and 7.30 show the sensitivity to profit of EVA with a varying number of EVs in V2G and G2V modes respectively. For a smaller group, profit is lower. For a larger group, the profit is considerably higher. There is a 7.44 % reduction in profit from risk-neutral to risk-averse in V2G mode as compared to a 1.33% reduction in profit from risk-neutral to risk-averse in G2V mode for 1000 EVs.

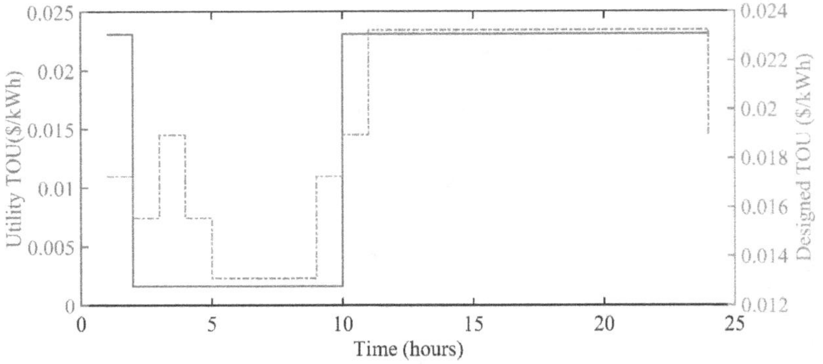

Fig. 7.26: Different TOU structures from utility and designed from EVA

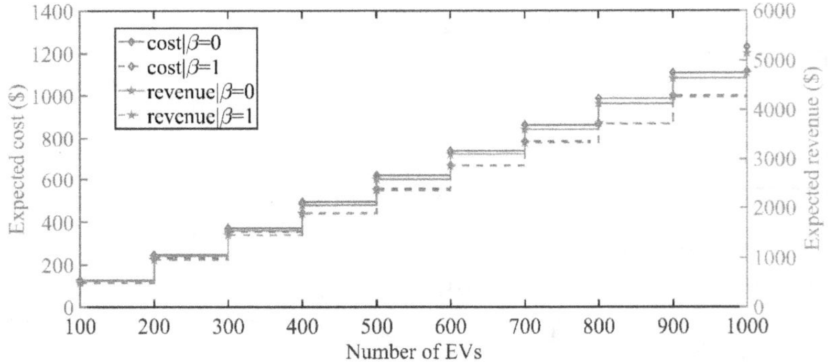

Fig. 7.27: Sensitivity of cost and revenue of EVA with Number of EVs in V2G

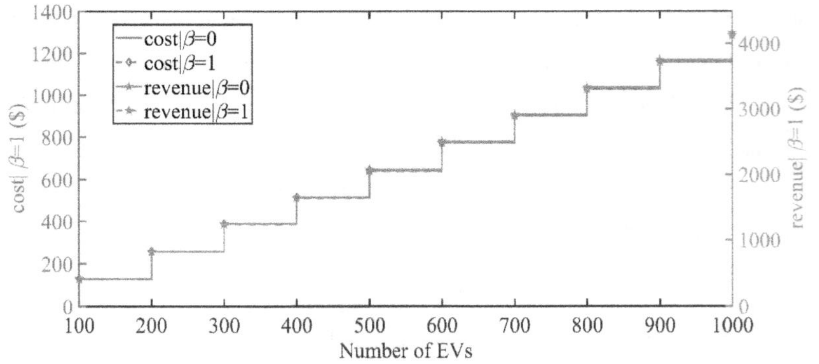

Fig. 7.28: Sensitivity to cost and revenue of EVA with a varying number of EVs in G2V

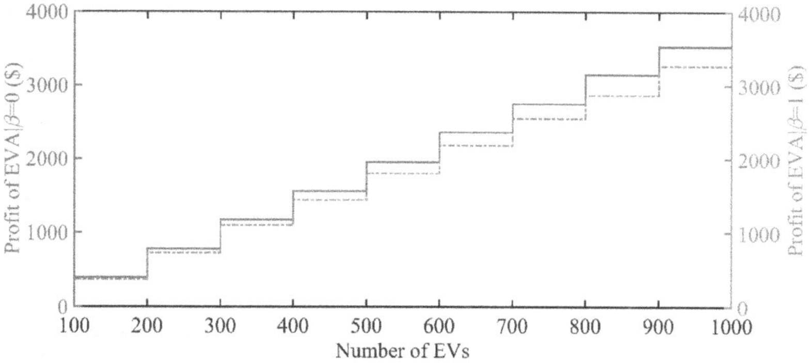

Fig. 7.29: Sensitivity of profit of EVA with Number of EVs in V2G

Figs. 7.31 and 7.32 show the Sensitivity of performance parameters of EVA with an upper limit of charging/discharging rate in V2G for RNDM and RADM. Figs. 7.33 and 7.34 show the Sensitivity of performance parameters of EVA with an upper limit of charging/discharging rate in G2V for RNDM and RADM. For a lower value of charging/discharging rate, profit is lower due to higher cost and lower revenue. For a higher value of charging/discharging rate, the profit is considerably higher due to lower cost and higher revenue. This is the integrated impact of TOU-PBDR from EV owners. There is a 22.2% reduction in cost, 33.82% reduction in revenue, and 14.92% reduction in profit from risk-neutral to risk-averse in V2G mode as compared to a 2.04 % reduction in cost, 3.38% reduction in revenue, and 5.61% increase in profit from risk-neutral to risk-averse in G2V mode for charging/discharging rate value 12.1 kW.

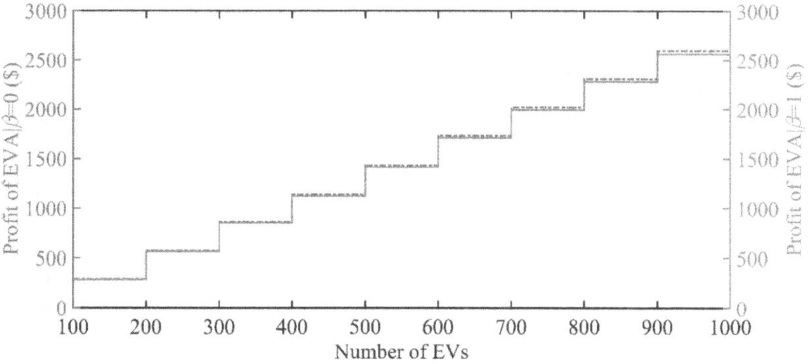

Fig. 7.30: Sensitivity of profit of EVA with Number of EVs in G2V

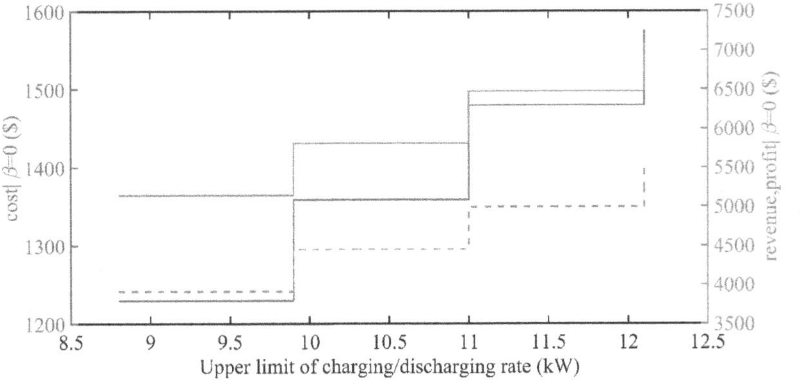

Fig. 7.31: Sensitivity of performance parameters of EVA with an upper limit of charging/discharging rate in V2G for RNDM

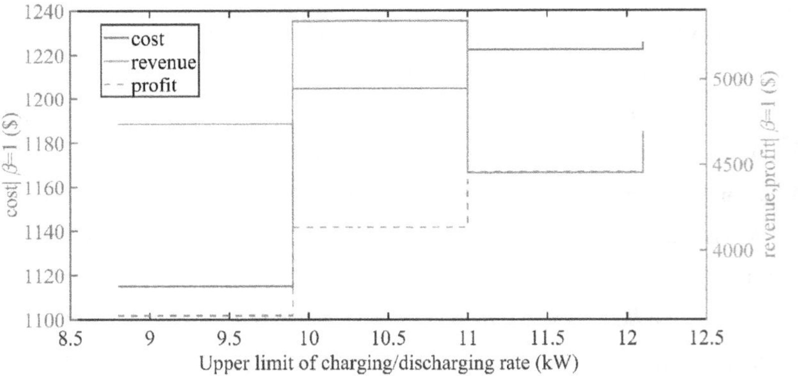

Fig. 7.32: Sensitivity of performance parameters of EVA with an upper limit of charging/discharging rate in V2G for RADM

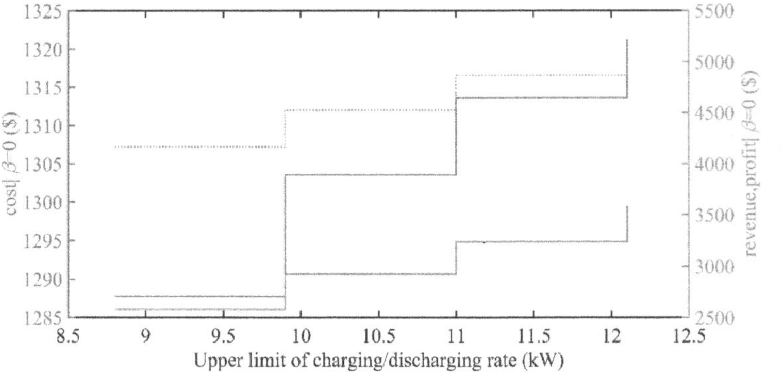

Fig. 7.33: Sensitivity of performance parameters of EVA with an upper limit of charging/discharging rate in G2V for RNDM

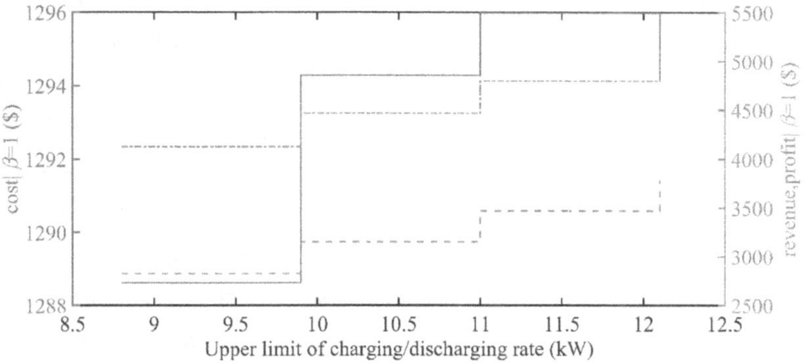

Fig. 7.34: Sensitivity of performance parameters of EVA with an upper limit of charging/discharging rate in G2V for RADM

7.5 Epilogue

The massive penetration of EVs poses threats to system security and reliability. Intuitively, EVA has great potential to participate in multi- markets. The frequent charge and discharge cycling of EV battery incurs faster depreciation and causes economic harms. DR integrated with coordinated V2G scheduling can address this issue. In this context, without loss of generality, optimal participation of EVA in energy and regulation markets has been investigated implementing TOU-PBDR from EV owners' perspective. EVA's profit varies due to market price uncertainty. Therefore, proposed work models risk-averse optimization framework for V2G management of EVs. *CVaR* risk measure is used for estimating the risks for managing the financial risk caused by uncertainties. EV owners must be compensated for participation in DRP due to battery degradation from the provision of additional services other than transportation. The degradation cost is integrated into the objective function of the proposed optimization. Also, a stochastic programming framework is used to model the uncertainty of the mobility behaviour and market price via scenarios. The proposed approach provides a range of benefits to all involved entities viz. EVA, EV Owners, and SO. This chapter proposes a novel exposition of the TOU price based V2G scheduling model for EVs coordinated charging/discharging operations. Simulation upon a cluster of 1000 traveller EVs from NYISO (the CAPITL Area) shows that V2G allows load shaping solution via energy storage in EVs, to support the grid during peak loads, or based on pricing signals. EVA can provide an insight into SO regarding the average amount of instantaneous curtailment capacity available (shed able load) during an emergency.

Through the EV market growth, the opportunity for demand-side management solutions persists to rise. DR captures easily system restraints and produces economic and security benefits for the market. EVA's optimal and prudent risk-aversive V2G (charging/discharging) scheduling incorporating TOU price-responsive DR for sale of grid-support services while ensuring fulfilment of EV owners' requirements using a stochastic programming framework can balance both the economic and technical objectives of EV owners and SO respectively. EVA supports the widespread adoption of EVs while benefiting the grid. Undue stress on distribution transformer can be avoided by the proposed scalable scheduling strategy. A key feature of this program is ensuring service provision would not negatively affect the DRP participants, thus could persuade EV owners to remain available for pro-active participation.

The proposed model simultaneously optimizes EV owners charging cost and revenue of EVA. Charging/discharging rates variation is done by EVA to decrease the charging load of EVs to maintain system operational limits. Discharging of EVs having higher SOC is performed at peak times. Results highlight the effectiveness of TOU pricing on EVA's strategy regarding charging/discharging rate variation and regulation capacities provided. Moreover, TOU pricing emboldens EV owners' participation in V2G regulation services. Outcomes of the proposed model could assist EVA or utility to derive V2G interaction of EVs for peak load management. The results demonstrated that existing distribution infrastructure could support large levels of EV penetration without significant overloading on the local grid due to discharged energy injected into the grid during peak hours.

The obtained results show that consideration of the degradation model decreases the obtained profit by about 33.78% in the short-term run. Furthermore, a risk management technique, using coherent risk measure *CVaR* is executed and compared with a risk-neutral problem. The expected profit for risk-neutral, RTP-DR, utility designed TOU-DR and designed TOU-DR are compared. Considering TOU-PBDR from a customer perspective, the expected profit in designed TOU-DR is obtained closer to RTP-DR whereas utility TOU-DR reduced the expected profit of EVA meaningfully. The risk-averse leads to a more conservatism condition compared to risk-neutral case, and the traded energy is reduced significantly. As a part of future scope risk-averse V2G scheduling at rooftop solar charging parking lot integrating PBDR and EV Owners' perspective can be formulated.

www.ingramcontent.com/pod-product-compliance
Lightning Source LLC
LaVergne TN
LVHW010726061126
851857LV00014B/1363
* 9 7 9 8 2 5 6 2 7 0 9 7 1 *